Collins · *do brilliantly!*

Instant**Facts**

Biology

A-Z of **essential facts** and definitions

T. A. McCahill

William Collins' dream of knowledge for all began with the publication of his first book in 1819. A self-educated mill worker, he not only enriched millions of lives, but also founded a flourishing publishing house. Today, staying true to this spirit, Collins books are packed with inspiration, innovation and practical expertise. They place you at the centre of a world of possibility and give you exactly what you need to explore it.

Collins. Do more.

Published by Collins
An imprint of HarperCollins*Publishers*
77–85 Fulham Palace Road
Hammersmith
London
W6 8JB

Browse the complete Collins catalogue at

www.collinseducation.com
© HarperCollins*Publishers* Limited 2005

First published as Collins Gem Basic Facts Biology 1996

10 9 8 7 6 5 4 3 2 1

ISBN 0 00 720511 2

British Library Cataloguing in Publication Data
A catalogue record for this publication is available from the British Library

Every effort has been made to contact the holders if copyright material, but if any have been inadvertently overlooked, the Publishers will be pleased to make the necessary arrangements at the first opportunity.

Edited and Project Managed by Marie Insall
Production by Katie Butler
Design by Sally Boothroyd/Jerry Fowler
Printed and bound by Printing Express, Hong Kong

You might also like to visit
www.harpercollins.co.uk
The book lover's website

Introduction

Instant Facts Biology is one of a series of illustrated A–Z subject reference guides of the key terms and concepts used in the most important school subjects. With its alphabetical arrangement, the book is designed for quick reference to explain the meaning of words used in the subject and so is an excellent companion both to course work and during revision.

Bold words in an entry identify key terms which are explained in greater detail in entries of their own; important terms that do not have separate entries are shown in *italic* and are explained in the entry in which they occur.

Other titles in the *Instant Facts* series include:
English
Modern World History
Science
Physics
Geography
Maths
Chemistry

A

abdomen **1.** (in mammals) The part of the body separated from the **thorax** by the diaphragm, containing **stomach**, **liver**, **intestines**, etc.
2. (in insects) The posterior third region of the body.

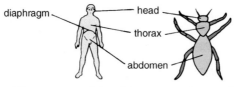

abdomen *The position of the abdomen in a human being and an insect.*

abiotic factor *See* **environment**.

abscission The shedding of leaves, fruit, and unfertilized flowers from plants by the formation of a layer of cork cells, which seal the plant surface and eventually cut off food and water from the part to be shed.

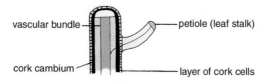

abcission *Cork cells seal off a leaf.*

absorption (of food) The process by which digested food particles pass from the gut into the bloodstream. In mammals absorption occurs in the **ileum**.

accommodation The ability of the eye of mammals to change its sharp focus from near to distant objects, and vice versa, by means of contraction or relaxation of the **ciliary muscles**, so altering the shape and hence the focusing properties of the **lens**.

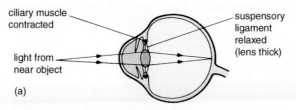

accommodation *(a) Eye focused on near object.*

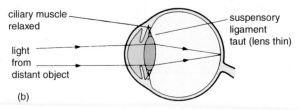

ciliary muscle relaxed

suspensory ligament taut (lens thin)

light from distant object

(b)

accommodation *(b) Eye focused on distant object.*

acid A substance that releases hydrogen **ions** in water and has a **pH** less than 7. An example is hydrochloric acid in the human stomach.

acid rain Rain that contains dissolved **acid** and is therefore a source of **pollution**. The acidity is caused mainly by the release of sulphur dioxide from burning **fossil fuels**. The sulphur dioxide reacts with water in the atmosphere to form sulphuric acid, which corrodes buildings, lowers the **pH** of soil and lakes, and affects plant and animal life.

active transport The movement of materials against a *concentration gradient* (*see* **diffusion**) using metabolic energy. Examples are (a) the uptake of **mineral salts** from soil by plant **root hairs** and (b) the reabsorption of certain substances by the mammalian kidney.

adaptation The development of structures within organisms so that they are more efficiently adapted to their environment. For example, plant leaves are constructed in such a way as to contribute to the efficiency of photosynthesis.
   Leaves are:
(a) thin, allowing rapid **diffusion** of air, thus facilitating **gas exchange**;
(b) flat and present a large surface area to the light.

ADH *See* **antidiuretic hormone.**

adipose tissue Mammalian tissue consisting of fat storage cells, located under the skin, around the kidneys, etc.

adolescence The period in the human **life cycle** between **puberty** and maturity.

ADP *See* **ATP.**

adrenal glands A pair of **endocrine glands** situated **anterior** to the mammalian kidneys and secreting the hormone *adrenalin*, which causes increased heartbeat, breathing, etc., in response to conditions of stress.

aerobe An organism that requires oxygen in order to survive. *See also* **anaerobe, respiration**.

aerobic bacteria *See* **sewage disposal**.

agar A jelly obtained from seaweed that is used as a medium for culturing bacteria. A food source, such as **glucose**, is added and the liquid agar is poured into a Petri dish where it solidifies. Bacterial sources are added to this *nutrient agar plate* and after two days at a suitable temperature (usually 37 °C), the bacterial colonies become visible as a result of rapid cell division. *See diagram.*

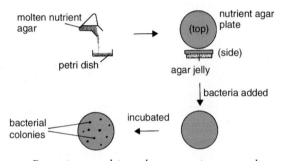

agar *Bacteria are cultivated on a nutrient agar plate.*

agglutination The process by which **red blood cells** clump together when the *antigens* on their surfaces react with complementary **antibodies**. *See also* **blood groups, blood transfusion**.

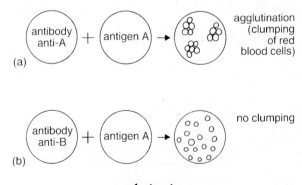

agglutination

agriculture The practice of farming, involving the cultivation of soil, crop production and raising livestock.

AIDS (aquired immune deficiency syndrome) A viral condition that weakens the body's immune responses. This results in reduced resistance to diseases such as pneumonia and can be fatal. The virus enters the bloodstream as a result of using contaminated needles, by transfusion of infected blood or blood-products, or via sexual contact with an infected person.

alcohol abuse The habitual excessive consumption of alcoholic drinks, which can lead to diseases of the liver, **nervous system** and **alimentary canal**.

algae A photosynthetic plant group including microscopic types such as *Spirogyra* and *Euglena*, and also **multicellular** types, e.g. seaweeds. Algae are widely distributed as marine and freshwater **plankton**, while some seaweeds are edible and others are a source of **agar**.

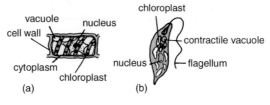

algae *(a) Spirogira. (b) Euglena.*

alimentary canal The digestive canal in animals. In humans it is a tube about nine metres in length running from mouth to **anus**. *See* **digestion**.

alkali An aqueous solution of a **base**, which releases hydroxyl ions in water and has a **pH** greater than 7. An example is lime (calcium hydroxide), which is added to soil to neutralize excess **acid**.

alleles Each inherited feature in an individual is produced by a pair of **genes** or alleles, which are located on the **chromosomes**. The pair of genes may produce the same or differing effects. For example, the fruit fly *Drosophila* has a pair of alleles that control wing length. The combination of these dictates the possession of normal wings or vestigial wings. *See also* **backcross**, **codominance**, **incomplete dominance**, **monohybrid inheritance**.

alveoli Air sacs in the mammalian **lungs** across which **gas exchange** occurs. *See also* **gas exchange** (*mammals*).

amino acids Organic compounds that are the subunits of **proteins**. Altogether some 70 different amino acids are known, but only about 20–24 are actually found in living organisms, bonded together in chains known as **peptides**, which are the basis of protein structure. *See* figure opposite.

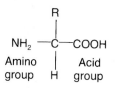

amino acid *Structure. (R is a variable group, depending on amino acid.)*

amniocentesis Removal of some fluid from the **amnion** during pregnancy. The fluid contains cells from the **foetus**, and by studying their **chromosomes** abnormalities such as **Down's syndrome** or **inherited diseases** can be detected.

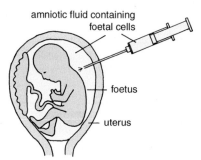

amniocentesis

amnion The fluid-filled sac surrounding and protecting the **embryos** of mammals, birds and reptiles. *See also* **pregnancy**.

anabolism *See* **metabolism**.

anaerobe An organism that lives in the absence of oxygen. *See also* **aerobe**, **respiration**.

annual A flowering plant that completes its **life history** from **germination** to death in one season.

antagonistic muscles Pairs of **muscles** which produce opposite movements, the contraction of one stimulating the relaxation of the other.

For example, at a joint, the contraction of the *flexor* (muscle that bends limb) stimulates the relaxation of the *extensor* (muscle that straightens limb) so that bending occurs. When the joint straightens due to contraction of the extensor, this causes the flexor to relax.

The longitudinal and circular muscles of the vertebrate gut and annelid worms also act antagonistically, the former causing **peristalsis**, and the latter movement. *See* figure overleaf.

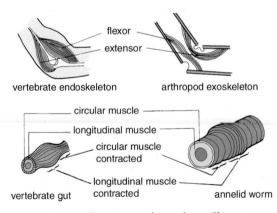

antagonistic muscles *Pairs of muscles pulling in opposite directions perform various functions.*

anterior Relating to parts of the body at or near the leading or head end of an animal. Opposite of **posterior**.

anther The upper part of a **stamen** in a flower. The anther contains **pollen** grains.

antibiotics Substances formed by certain bacteria and fungi that inhibit the growth of other microorganisms. For example, **penicillin** and streptomycin.

antibiotic discs Sterile paper discs containing **antibiotic**. They are used to identify which antibiotic will be effective against infection with a particular bacterium. The discs are added to nutrient **agar** plates that have been contaminated with bacteria taken from a patient.

In plate A, bacteria grow only around the **penicillin** disc, while the plate is clear around the *streptomycin* disc. This shows that not all antibiotics are effective against all microorganisms. This is confirmed by plate B, in which three of the antibiotics had no effect.

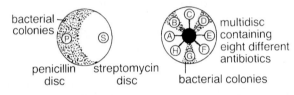

antibiotic discs *Possible results after incubation. Certain antibiotics have destroyed the bacteria.*

antibiotic resistance The ability of certain microorganisms to overcome the action of **antibiotics**: this results from **mutations** producing new strains that are no longer susceptible to the antibiotic.

antibodies Proteins produced by vertebrate tissues as a reaction to **antigens**, i.e. materials foreign to the organism (for example, microorganisms such as bacteria and their **toxins**, or transplanted organs or tissues). *See also* **agglutination**.

antigens   antibodies   antigens neutralized

antibodies *Antibodies react with antigens to make them harmless.*

antidiuretic hormone (ADH) A **hormone** secreted in mammals by the **pituitary gland**. ADH stimulates water reabsorption by the kidneys, thus reducing water loss in the urine.

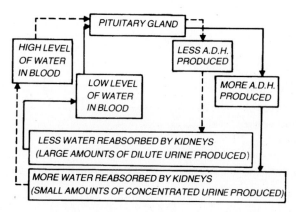

antidiuretic hormone *Control of water loss.*

antigen A substance, usually a protein, that stimulates the immune system to make an **antibody**. *See* **immunity**.

antiserum (*pl.* **antisera**) A substance containing large numbers of **antibodies** to a particular **antigen**. *See* **blood tests**.

anus The terminal opening of the **alimentary canal** in mammals, through which **faeces** are shed. The anus is opened and closed by a **sphincter muscle**.

aorta The largest **artery** in the mammalian **circulatory system**; it carries blood from the left **ventricle** of the heart to the rest of the body.

appendix A small sac found at the junction of the **ileum** and **caecum** of some mammals. In humans its function is unclear.

aqueous humour Clear watery fluid filling the front chamber of the vertebrate eye between the **cornea** and the **lens**.

artery A vessel that transports blood from the heart to the tissues. In mammals, arteries carry *oxygenated* blood, i.e. blood which carries oxygen in the haemoglobin, around the

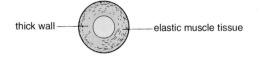

thick wall — elastic muscle tissue

artery *Section through an artery.*

body (for an exception to this rule, *see* **pulmonary vessels**) and divide into smaller vessels called *arterioles*. Arteries have thick, elastic, muscular walls, in order to withstand the high pressure caused by the heartbeat.

arterioles *See* **artery**.

artificial propagation The method by which plant growers make use of a plant's capacity for **asexual reproduction** and **regeneration** in order to produce new plants. Small pieces of stem, root or leaf are placed into suitable conditions to encourage the growth of new plants (**clones**). Three methods are shown here.

artificial selection The method by which animal and plant breeders attempt to improve stocks by selecting males and females with desirable

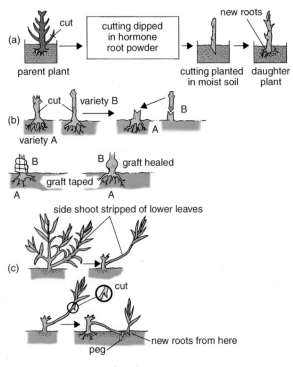

artificial propagation
(a) Cutting. (b) Grafting. (c) Layering.

characteristics and allowing them to interbreed. By rigorous selection over many generations, improvements can be made in stock quality, e.g. in beef and milk production in cattle or yield and disease resistance in crops. *See also* **natural selection**.

asexual reproduction **Reproduction** in which new organisms are formed from a single parent without **gamete** production. The offspring from asexual reproduction are genetically identical to each other and to the parent organism, and are referred to as **clones**. *See also* **binary fission, budding, spore, vegetative reproduction**.

assimilation (of food) The process by which digested food particles are incorporated into the **protoplasm** of an organism. For example, in mammals glucose not required immediately to provide energy in tissue respiration is converted in liver and muscle cells to the storage carbohydrate, **glycogen**, which can be reconverted to glucose if the blood glucose level falls (*see* **insulin**). Surplus glucose not stored as glycogen is converted to fat and stored in fat storage cells beneath the skin, as a long-term energy store.

Fatty acids and glycerol (*see* **fats**) are reassembled into fat. Surplus fat is stored as outlined above.

Amino acids are synthesized into **proteins**. Surplus amino acids cannot be stored and are disposed of in the liver by **deamination**.

atom The smallest complete particle of an **element** that can exist chemically. Each atom consists of a nucleus of **protons** and **neutrons** surrounded by orbiting **electrons**. *See also* **molecule**.

ATP (adenosine triphosphate) A chemical compound that acts as a store and a source of energy within **cells**. ATP is formed from *adenosine diphosphate* (ADP) and a phosphate group using energy from **respiration**, which can then be released for metabolic processes when ATP is broken down.

ATP *ATP provides energy for metabolic processes.*

atrium *See* **heart, heartbeat**.

auditory Relating to part of the body and functions connected with the **ear**.

auditory canal A tube in the mammalian outer ear leading from the **pinna** to the **tympanum**.

auditory nerve A **cranial** nerve in vertebrates that conducts **nerve impulses** from the inner ear to the brain. *See also* **ear**.

auricle *See* **heart**, **heartbeat**.

autoclave A pressure cooker used for the **sterilization** of materials, such as **agar** before and after use in **microbiology** experiments. The materials to be sterilized are heated in the autoclave at 120 °C for 15 minutes to destroy any bacteria present.

autoradiograph A picture obtained when a photographic negative is exposed to living tissue into which radioactive material has been introduced in order to trace the route of substances through the tissue.

autotrophic (used of organisms) Able to synthesize complex **organic compounds** from simple non-living **inorganic compounds**. The major autotrophs are green plants, which use water and carbon dioxide to make food by photosynthesis. For this reason green plants are also called **food producers**. *Compare* **heterotrophic**.

auxins **Plant growth substances** that control many aspects of plant growth, for example, **tropisms**, by stimulating **cell division** and elongation.

axon *See* **neurones**, **synapse**.

B

backbone *See* **vertebral column**.

backcross A **genetic** cross in which a **heterozygous** organism is crossed with one of its **homozygous** parents. Thus two backcrosses are possible.

For example, in the fruit fly *Drosophila*, normal wings are **dominant** to vestigial wings and so heterozygous flies will have normal wings. The backcross with the **recessive** homozygote is useful in distinguishing between organisms with the same **phenotype** but different **genotypes**. Examples are NN and Nn. Such a cross is called a testcross.

See also **monohybrid inheritance**.

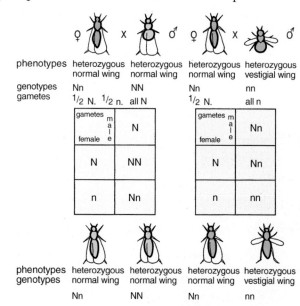

backcross *The two backcrosses of Drosophila.*

bacteria **Unicellular** organisms with a diameter of 1–2 microns. Some bacteria cause disease such as tetanus, but others are useful as, for example, sources of **antibiotics**.

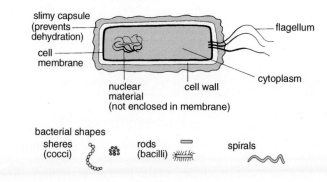

bacteria *Structure of a generalized bacterium, with bacterial shapes.*

Baermann funnel An apparatus used to isolate organisms living in soil water, such as **algae** and **protozoa**. The organisms move away from the strong light and high temperature of the lamp and collect near the tap, from where they can be released into the collecting jar.

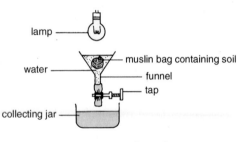

Baermann funnel

balanced diet The correct nutritional components required for health, generally used in reference to human beings and domesticated animals. A balanced diet for humans should contain:

(a) Sufficient kilojoules of energy;
(b) Protein;
(c) Carbohydrate;
(d) Fat;
(e) Vitamins;
(f) Water;
(g) **Mineral salts**;
(h) **Roughage** (*fibre*).

basal metabolic rate (BMR) The rate of **metabolism** of a resting animal as measured by oxygen consumption. BMR is the minimum amount of energy needed to maintain life and varies with species, age and sex.

base A compound, usually a metallic oxide or hydroxide, which, if soluble in water, forms an **alkali**.

base pair *See* **genes**.

batch processing (in **biotechnology**) An industrial process in which **enzymes** are dispersed throughout a **substrate**. At the end of the batch, the enzyme has to be separated from the product, and the **fermentation** vessel must be emptied and cleaned before the next batch. *See also* **continuous flow processing**.

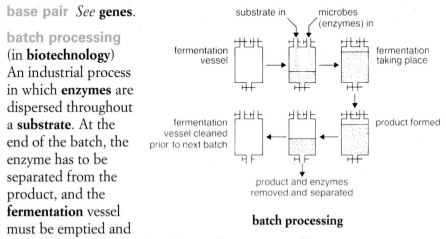

batch processing

bile A green alkaline fluid produced in the **liver** of mammals. Bile is stored in the **gall bladder** and is transported via the *bile duct* to the

duodenum where it causes fat to be broken into minute droplets (*emulsification*) before digestion.

binary fission **Asexual reproduction** in **unicellular** organisms in which a single cell divides to produce two cells. The nucleus divides by **mitosis**. Binary fission is common in bacteria and **protozoa** such as *Amoeba* where a single mother cell divides into two identical daughter cells.

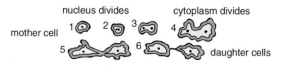

binary fission *Amoeba.*

binomial nomenclature The method of naming organisms devised by Carl Von Linne (Linnaeus) in the 18th century. Each organism has two Latin names, the first, with an initial capital, indicating the **genus**, and the second with a lower case first letter indicating **species**. For example:

Genus	Species	Common name
Canis	*familiaris*	domestic dog
Canis	*lupus*	American wolf

See also **classification**.

biodegradable Able to be readily broken down by biological action. For example, biodegradable detergents in sewage can be digested by bacteria.

biological clock The mechanism thought to be responsible for **rhythmical behaviour** patterns in animals. Also associated with repeating natural cycles such as tides, **photoperiods** and seasons.

biological control The use of natural **predators**, **pathogens** and **parasites** to control **pests**.

biological detergents Powders containing enzymes obtained from bacteria. These enzymes break down stains caused by proteins in milk, blood, egg etc. The stains are converted by the enzymes into soluble substances that can be washed away.

biomass The total mass of living matter in a **population**. Biomass is usually expressed as living or dry weight and decreases at each level in a **food chain**. *See also* **pyramid of biomass, pyramid of numbers**.

biosensor A device used for the rapid detection of chemicals in blood or urine. A biosensor uses **immobilization** to detect a particular substance by reacting specifically with a substance to give a product. This is then used to generate an electrical signal. An example of a biosensor is the *glucose oxidase electrode*, which measures the amount of glucose in blood.

biosphere The part of the Earth that contains living organisms. The biosphere includes all the various **habitats** from the deepest oceans to the highest mountains. *See also* **environment**.

biotechnology The use of living organisms in manufacturing processes. For example, microorganisms are used in ethanol production by **fermentation**, **brewing** with **yeast**, and **enzyme** production.

biotic factor *See* **environment**.

birth (in humans) The process by which a **foetus** leaves the **uterus** to live outside the mother as an individual being. The human baby is born as a result of muscular contractions of the uterus wall. The *amniotic fluid*, in which the baby has been floating, escapes, and the baby is pushed through the **cervix** and the **vagina** and thus leaves the mother's body.

   The **umbilical cord** is cut, the **placenta** is expelled as the *afterbirth* and the baby must now use its own **lungs** for **gas exchange**. *See also* **fertilization**, **pregnancy**.

birth rate (of a **population**) The number of live births, measured in the human population as number of births in one year per 1000 of population. *See also* **human population curve**.

bladder (urinary) A sac into which **urine** from the kidneys passes via the **ureters**. From the bladder, urine is discharged through the **urethra**. *See* **kidney**.

blind spot The area of the **retina** where the **optic nerve** passes out of the eyeball, and which is devoid of **rods** or **cones**. The blind spot can be demonstrated by a simple experiment, as follows. Hold the book at arm's length. Close the left eye and concentrate on the cross with the right. Slowly bring the book closer until the drawing of the face seems to disappear. At this point the image of the face is falling on the blind spot.

blind spot

blood A fluid tissue found in many animals with the principal function of transporting substances from one part of the body to another. In mammals, blood consists of a watery solution called **plasma**, in which there are three types of cells: **platelets, red blood cells** and **white blood cells.**

The main functions of blood are:

(a) Transport of oxygen from lungs to tissues;

(b) Transport of toxic by-products to the organs of **excretion**;

(c) Transport of hormones from **endocrine glands** to target organs;

(d) Transport of digested food from the **ileum** to the tissue;

(e) Prevention of infection by **blood clotting, phagocytosis** by white blood cells, and **antibody** production.

blood clotting The conversion of blood **plasma** into a clot, which occurs when blood **platelets** are exposed to air as a result of injury. The platelets produce an enzyme (*thrombin*) that causes the conversion of a soluble **plasma protein (fibrinogen)** into *fibrin*, which forms a meshwork of fibres. The resulting clot restricts blood loss and the entry of microorganisms.

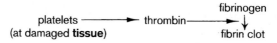

blood clotting *The process of clot formation.*

blood groups The classification of **blood** types based on the different antigens (*see* **antibodies**) present on the surface of **red blood cells.**

The human population is divided into four groups called A, B, AB and O. The capital letters stand for the type of antigen present in the red blood cells. The corresponding antibodies are carried in the **plasma**, and if a person has a particular antigen in his red cells, he cannot have the corresponding antibody since **agglutination** would occur. Thus:

Group A contains antigen A and antibody anti-B.

Group B contains antigen B and antibody anti-A.

Group AB contains antigens A and B and no antibodies of either type.

Group O contains no antigens and has antibodies anti-A and anti-B.

See also **blood transfusion.**

blood group	antigen on red blood cells	antibody in plasma
A	A	anti-B
B	B	anti-A
AB	A and B	neither
O	neither	anti-A and anti-B

blood pressure The pressure of **blood** in the main **arteries** of mammals. In humans, blood pressure is normally about 120 mm mercury (Hg) at systole (*see* **heartbeat**) and about 80 mm Hg at diastole, but can vary with age, exercise etc.

blood serum Fluid consisting of blood **plasma** with the **fibrinogen** removed.

blood tests The use of **agglutination** in order to determine a person's **blood group**. If one drop of blood from a sample is mixed with anti-A **blood serum** and another drop with anti-B blood serum then:
Group A red blood cells will clump only in anti-A;
Group B red blood cells will clump only in anti-B;
Group AB red blood cells will clump in both antisera;
Group O red blood cells will clump in neither antiserum.
See also **agglutination**, **blood groups**, **blood transfusions**.

blood group under test	antigens on cells	antibodies in plasma	added to anti-A test serum	added to anti-B test serum
A	A	anti-B		
B	B	anti-A		
AB	A and B	neither		
O	neither	anti-A anti-B		

blood tests *Summary.*

blood transfusion The transfer of **blood** from a healthy person (the *donor*) to another person who has lost a lot of blood, e.g. as a result of an injury. The **blood groups** of the donor and the patient must match or else the **antibodies** in the patient's **plasma** will act upon the **antigens** on the donor's **red blood cells** and cause the cells to clump together (**agglutination**). *See also* **blood groups**.

blood group	can donate blood to	can receive blood from
A	A and AB	A and O
B	B and AB	B and O
AB	AB	all groups
O	all groups	O

blood transfusion

blood vessels Tubes transporting **blood** around the bodies of many animals, which together with the **heart** make up the **circulatory system**. In vertebrates, the blood vessels consist of **arteries**, arterioles, **capillaries**, venules and **veins**.

BMR *See* **basal metabolic rate**.

bone Tissue in the vertebrate **skeleton** consisting of **collagen** (a protein), which gives **tensile strength**, and calcium phosphate, which gives bone its hardness. Some bones have a hollow cavity containing bone marrow in which new **red blood cells** are produced.

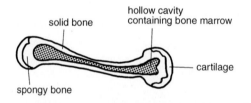

bone *Section through bone.*

Bowman's capsule A cup-shaped part of a **kidney** tubule or **nephron** in mammals.

brain The large mass of **neurones** in animals, which has a centralized coordinating function. In vertebrates it is found at the **anterior** end of the body, protected by the **cranium**, and connected to the body via the **spinal cord** and *spinal nerves*, and directly by nerves called *cranial nerves*, such as the **optic nerve**, **auditory nerve**.

The human brain contains millions of nerve cells, which are continually receiving and sending out **nerve impulses**. The remarkable property of the brain is that it translates electrical impulses in such a way that stimuli from the environment, such as sound and light, are appreciated so that the recipient of the stimuli can respond and adapt to the environment in the most appropriate way. The brain also coordinates bodily activities to ensure efficient operation, and stores information so that behaviour can be modified as the result of experience.

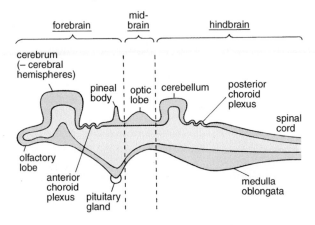

brain *Diagram of the vertebrate brain.*

breastbone *See* **sternum**.

breathing (in mammals) The inhalation and exhalation of air for the purpose of **gas exchange**. In mammals the gas exchange surface is situated in the **lungs**.

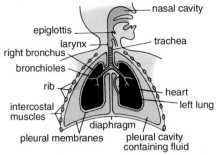

breathing *Lungs and associated structures.*

The exchange of air in the lungs (*ventilation*) is caused by changes in the volume of the **thorax**, brought about by the action of the **diaphragm** and **intercostal muscles**.

When the diaphragm contracts, it depresses, increasing the volume of the thorax, causing air to rush into the lungs. Relaxation of the diaphragm reduces the volume of the thorax, and causes exhalation of air. The action of the diaphragm is accompanied by the raising and lowering of the rib cage, which is necessary to accommodate the changes in lung volume. These rib cage movements are caused by contraction and relaxation of the intercostal muscles.

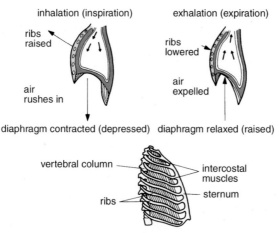

breathing *Movements of the rib cage.*

breathing rate The rate of **lung** ventilation. In humans, **breathing** movements are controlled by the **medulla oblongata** in the brain, which is sensitive to the carbon dioxide concentration of the blood. If the carbon dioxide concentration rises sharply as the result of increased **respiration**, for example, during exercise, the brain sends **nerve impulses** to the **diaphragm** and **intercostal muscles**, which react by increasing the rate and depth of breathing. This accelerated breathing rate helps to expel the excess carbon dioxide and increases the supply of oxygen to respiring cells.

brewing The manufacture of beer by **fermentation**. The main stages are as follows:

1) Germinating barley grains convert

    starch $\xrightarrow{\text{enzymes}}$ sugar (*malting*)

2) sugar + **yeast** + hops
    |
    fermentation
    ↓
    alcohol + carbon dioxide
    (beer)

Other alcoholic brews can be manufactured using **substrates** other than barley and hops, e.g. apple juice (to make cider), grape juice (wine), rice (*sake*) and honey (mead).

bronchus One of two air passages branching from the **trachea** in lunged vertebrates. *See* **lungs**.

budding **Asexual reproduction** in which a new organism develops as an outgrowth or *bud* from the parent; the offspring often becoming completely detached from the parent. Budding is common among coelenterates, e.g. *Hydra* and **unicellular fungi**, e.g. **yeast**. *See diagram.*

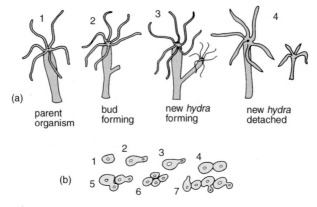

(a)
parent | bud | new *hydra* | new *hydra*
organism | forming | forming | detached

(b)

budding *(a) Hydra. (b) Yeast.*

buffer A **solution** that maintains a constant **pH** even on the addition of an **acid** or an **alkali**.

bulb The organ of **vegetative reproduction** in flowering plants, consisting of a modified **shoot** whose short stem is enclosed by fleshy scale-like leaves. In the growing season, one or more buds within the bulb develop into new plants, using food stored in the bulb.

Bulb-producing plants include the tulip, daffodil and onion.

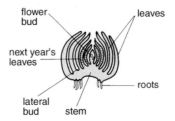

flower bud | leaves
next year's leaves
lateral bud | stem | roots

bulb *Section through bulb.*

C

caecum Part of the mammalian gut at the entry to the **large intestine**. In **herbivores** it is very important in **cellulose digestion**. In humans it is much reduced in size and its function is uncertain.

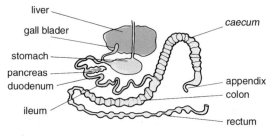

caecum *The digestive system of herbivores.*

canines Sharp-pointed tearing **teeth** near the front of the mouth used for killing prey, and ripping off pieces of food. Often reduced or missing in **herbivores**, present in **omnivores** and prominent in **carnivores**. *See also* **dental formula**, **dentition**.

capillaries Very small **blood vessels** that branch off from *arterioles* and form a network in vertebrate tissues, the blood eventually draining into *venules* and then **veins**.

Capillary walls are only one cell thick, allowing diffusion of substances between the blood and the tissues via a liquid called *tissue fluid* (**lymph**).

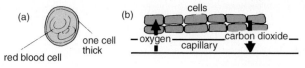

capillaries *(a) Section through capillary. (b) Capillary and tissues.*

carbohydrase Any **enzyme** that breaks down **carbohydrate**, by **hydrolysis**, into **disaccharides** and **monosaccharides**; examples are salivary amylase and maltase. *See* figure overleaf.

carbohydrates Organic compounds containing the elements carbon (C), hydrogen (H) and oxygen (O) and with the general formula CH_2O. Carbohydrates are either individual **sugar** units or chains of sugar units bonded together. The three main carbohydrate types are **monosaccharides**, **disaccharides** and **polysaccharides**.

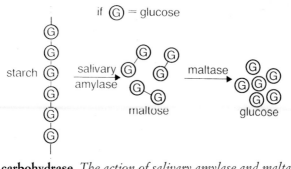

carbohydrase *The action of salivary amylase and maltase.*

The importance of carbohydrates is as follows:

(a) Simple carbohydrates, particularly **glucose**, are the principal **energy** source within cells;

(b) Long-chain carbohydrates form some structural cell components, for example, **cellulose**, in plant cell walls, and also act as food reserves, for example, **glycogen** in animals and **starch** in plants.

carbon cycle The circulation of the element carbon and its compounds, in nature, caused mainly by the **metabolism** of living organisms.

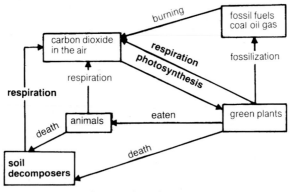

carbon cycle *The main steps.*

cardiac Relating to activities and parts of the body connected with the **heart** and its functions.

carnassials Shearing **teeth** used for cutting meat into chunks. These teeth are typical of **carnivores** and replace the premolars and molars found in **herbivores** and **omnivores**. *See also* **dental formula**, **dentition**.

carnivore An animal that feeds on flesh. Carnivores include dogs, cats, etc. They have a **dentition** adapted for killing prey, shearing raw flesh, and cracking bones. The outstanding features of carnivore dentition are the large piercing **canine teeth**, and the shearing **carnassial** teeth. The lower jaw can usually only move up and down, forming an effective clamp on the prey. Carnivores typically have two sets of teeth during their lives.

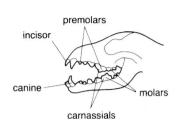

carnivore *The skull and teeth of the dog.*

carpel The female part of a **flower**. It contains an **ovary** in which there are varying numbers of **ovules**. These contain **embryo sacs**, within which are the female **gametes**. *See also* **fertilization**.

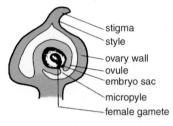

carpel *The female reproductive part of a flower.*

cartilage Supporting tissue found in vertebrates. In mammals there is cartilage in the larynx, trachea, bronchi and at the ends of bones at moveable joints, while in some fish, e.g. sharks, the entire skeleton is cartilage.

catabolism *See* **metabolism**.

catalyst A substance that accelerates a chemical reaction but remains unchanged at the end of the reation. *See also* **enzyme**.

caudal Relating to that part of the body of an animal at or near the tail.

cell A unit of **cytoplasm** governed by a single **nucleus** and surrounded by a **selectively permeable membrane**. Cells are the basic units of which most living things are made.

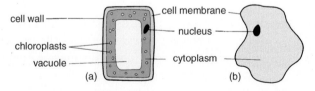

cell *Structure of (a) a plant cell and (b) an animal cell.*

The nucleus contains the hereditary material, the **chromosomes**, and controls the cell's activities. The cytoplasm is the liquid 'body' of the cell in which the chemical reactions of life occur, for example, **respiration**.

The *cell membrane* controls the entry and exit of materials, allowing certain substances through, but preventing the passage of others. Such a membrane is described as a **selectively permeable membrane**.

The *cell wall* is found only in plants. It is made of **cellulose** and gives shape and rigidity to the cells.

Chloroplasts are structures within green plant cells where **photosynthesis** occurs.

The **vacuole** is filled with cell sap in plants. The sap, when in sufficient quantity, creates a pressure on the cytoplasm and cell wall and helps keep the cell firm and resilient. *See also* **protoplasm**.

cell differentiation The process of change in **cells** during growth and development, whereby previously undifferentiated cells become specialized for a particular function as a result of structural changes.

For example in plant cells, following **cell division**, the daughter cells increase in size (*elongation*) by absorbing water. After elongation, cell differentiation occurs as the result of changes in the **protoplasm** and cell wall.

For example:
(a) Some cells have their walls strengthened by additional **cellulose**, e.g. **cortex**, **epidermis**.
(b) Some cells have **lignin** deposited in their walls, e.g. **xylem**.
(c) Some cells develop extra **organelles**, for example, the **palisade mesophyll** cells of leaves develop a high number of **chloroplasts**.

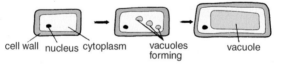

cell wall nucleus cytoplasm   vacuoles   vacuole
                                     forming

cell differentiation *Elongation in plant cells.*

cell division The division of the **cell** and its contents into two. The nucleus usually divides by **mitosis**, a process that gives the **nuclei** of the *daughter cells* exactly the same number of **chromosomes** as the *mother nucleus*. When a cell divides to form **gametes** (sex cells) the nucleus divides by **meiosis**, which provides daughter nuclei with half the original number of chromosomes. In animals the **cytoplasm** divides by constriction into two. In plants a wall is laid down between the two halves.

cell membrane *See* **cell**, **selectively permeable membrane**.

cellulose The **polysaccharide** carbohydrate that forms the framework and gives strength to plant cell walls. Cellulose remains undigested in the human gut but has an important role as **roughage**. In mammalian **herbivores**, in the **caecum** and **appendix**, bacterial populations produce an enzyme called *cellulase*, which digests cellulose.

cellulose *The breakdown of cellulose into glucose by the action of cellulase.*

cell wall *See* **cell**.

central nervous system (CNS) The part of the vertebrate **nervous system** that has the highest concentration of **neurone** bodies and **synapses**, i.e. the **brain** and **spinal cord**.

cerebellum The region of the vertebrate **brain** that in mammals controls balance and muscular coordination, allowing precise control during activities such as walking and running.

cerebrum or cerebral hemispheres The region of vertebrate **brain** that in mammals makes up the largest part of the brain. In humans the cerebrum consists of right and left hemispheres, the outer part made up of **neurone cell bodies** (*grey matter*), the inner part consisting of nerve fibres (*white matter*). The human cerebrum is responsible for the higher mental skills such as memory, thought, reasoning and intelligence. The cerebrum also contains localized areas concerned with specific functions. Areas receiving **nerve impulses** from **receptors** are called *sensory areas*, while those sending out impulses to **effectors** are called *motor areas*.

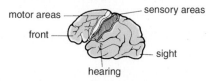

cerebrum *Left cerebral hemisphere of human brain, showing localization of functions.*

cervical Describes parts of the body and functions related to (a) the neck, (b) the **cervix**.

cervix The posterior region of the mammalian **uterus**, leading into the **vagina**. *See also* **fertilization**.

chemoreceptor A **receptor** that is stimulated by chemical substances. Examples are smell and taste receptors.

chemotropism **Tropism** relative to chemical substances. The growth of **pollen** tubes towards the **ovary** is an example of positive chemotropism (*see* **fertilization**).

chlorophyll Green pigment found in the **chloroplasts** of plant cells, which can absorb the light energy required for **photosynthesis**. The importance of chlorophyll can be shown by the *variegated leaf test* for **starch**. Only those parts of the leaf that were previously green (i.e. that contained chloroplasts) give a positive starch test, showing that chlorophyll is necessary for photosynthesis.

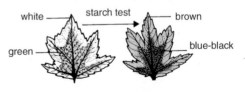

chlorophyll *The variegated-leaf test.*

chloroplasts Structures in the **cytoplasm** of green plant **cells**, in which **photosynthesis** occurs. Chloroplasts contain the green pigment **chlorophyll**. *See also* **palisade mesophyll**.

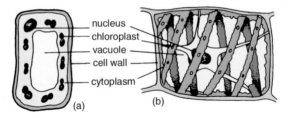

chloroplasts *The chloroplasts of (a) a palisade mesophyll cell and (b) the alga Spirogyra.*

choroid A layer of cells outside the **retina** of the vertebrate eye.

chromosomes The hereditary material within the **nucleus** of **cells**, which links one generation with the next. Each species has characteristic numbers and types of chromosomes.

For example, in humans, the chromosome number is 46. When a nucleus divides by **mitosis**, this **diploid** number of chromosomes is maintained in the newly formed nuclei. **Haploid** nuclei contain half the diploid number of chromosomes, and are made when the nucleus divides by **meiosis**. Two haploid **gametes** join to form a diploid **zygote**.

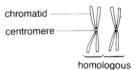

chromosomes *A pair of homologous chromosomes.*

Chromosomes control cellular activity. They consist of sub-units called **genes**, which contain coded information in the form of the chemical compound **DNA**. In diploid cells, chromosomes occur in similar pairs known as homologous pairs. Thus a human diploid cell contains 23 pairs of **homologous chromosomes**.

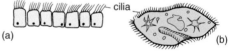

cilia Microscopic motile threads projecting from certain cell surfaces that stroke rhythmically together like oars. Cilia occur in certain vertebrate **epithelia** where

cilia *The cilia of (a) epithelial cells and (b) Paramecium.*

they cause movement of particles in the **trachea, oviduct, uterus,** etc. In some **protozoa,** for example *Paramecium,* cilia cause movement of the whole organism. *Compare* **flagellum.**

ciliary muscle Tissue in the vertebrate eye responsible for **accommodation**.

circulatory system Any system of vessels in animals through which fluids circulate, e.g. the blood circulation, **lymphatic system**. In mammals there are two overlapping blood circulations, i.e. there is a circulation between heart and lungs and a circulation between heart and body. This arrangement is called a *double circulatory system.* Blood flows through both circulations, always in the same direction, passing repeatedly through the heart.

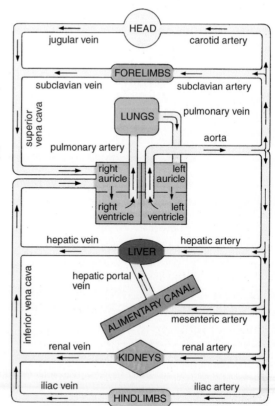

circulatory system *The double circulatory system of mammals.*

class A unit used in the **classification** of living organisms, consisting of one or more **orders**.

classification The method of arranging living organisms on the basis of similarity of structure into groups that show how closely they are related to each other and also indicate evolutionary relationships.

  The modern system of classification was devised by Carl von Linne (Linnaeus) in the eighteenth century. Organisms are first sorted into large groups called **kingdoms**, which are divided into smaller groups called **phyla** in animals and **divisions** in plants, then **classes**, **orders** and **families**, each subdivision producing subsets containing fewer and fewer organisms, but with more and more common features. Ultimately organisms are grouped in genera (singular **genus**), which are groups of closely related **species**. It is not uncommon for scientists to disagree as to how to classify certain organisms. *See table below.*

	Human	**Dog**	**Oak**	**Meadow buttercup**
Kingdom	Animal	Animal	Plant	Plant
Phylum/Order	Chordata	Chordata	Spermatophyta	Spermatophyta
Class	Mammalia	Mammalia	Angiospermae	Angiospermae
Order	Primates	Carnivora	Fagales	Ranales
Family	Hominidae	Canidae	Fagaceae	Ranunculaceae
Genus	*Homo*	*Cani*	*Quercus*	*Ranunculus*
Species	*sapiens*	*familiaris*	*robur*	*acris*

classification *How four organisms are classified.*

clavicle or collarbone The **ventral bone** of the shoulder-girdle of many vertebrates. It articulates with the **scapula** and **sternum**. *See* **endoskeleton**.

cloaca The posterior region of the **alimentary canal** in most vertebrates (but excluding mammals) into which the terminal part of the intestine, the kidney and reproductive ducts open.

clone A group of organisms that are genetically identical to each other, having been produced by **asexual reproduction**.

cloning methods *See* **artificial propagation**.

cochlea A spiral structure in the mammalian inner **ear** containing an area called the *organ of Corti* in which are located **neurones** that are sensitive to sound vibrations.

codominance A situation in which both **alleles** are expressed equally in the **phenotype** of a **heterozygote**. The inheritance of human **blood groups** includes an example of codominance. There are four blood group phenotypes: A, B, AB and O. The **genes** for groups A and B are codominant and both are completely dominant to the gene for group O. Thus if a person inherits genes for group A and group B, half his **red blood cells** will carry antigen A and half antigen B (*see* **antibodies**).

Phenotype (blood group)	genotype
A	$I^A I^A$ or $I^A I^O$
B	$I^B I^B$ or $I^B I^O$
AB	$I^A I^B$
O	$I^O I^O$

codominance *The phenotypes and genotypes of the four blood groups using the symbol I to represent the alleles.*

CNS *See* **central nervous system.**

cold blooded *See* **ectotherm.**

collagen Fibrous protein, which is the principal component of vertebrate **connective tissue**, and an important skeletal substance in higher animals, conferring **tensile strength** to bones, tendons and ligaments.

collarbone *See* **clavicle.**

colon A region of the **large intestine** in mammals between the **caecum** and **rectum**, which receives undigested food from the **ileum**. In the colon, much of the water is absorbed from the undigested food, and the semi-solid remains (**faeces**) are passed into the **rectum**. *See also* **digestion.**

commensalism A symbiotic relationship in which one organism benefits, while the other neither suffers nor benefits. For example, a marine worm lives in a shell with a crab, sharing the crab's food, but giving nothing in return. *See also* **symbiosis.**

community The **population** of different **species** living in a particular **habitat** and interacting with each other. For example, a rockpool habitat may have a community made up of crabs, worms, sponges, seaweeds, etc. *See also* **niche.**

companion cells Cells in flowering plants, that are associated with **phloem** *sieve tubes*, and are believed to contribute to the transport function of phloem (**translocation**).

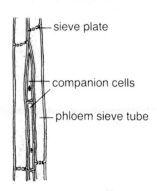

companion cell

compensation point (of green plants) The light intensity at which the rate of carbon dioxide uptake (**photosynthesis**) is exactly equal to the rate of carbon dioxide production (**respiration**). In a single day there are two compensation points when the rate of photosynthesis (**carbohydrate** gain) is exactly balanced by the rate of respiration (carbohydrate loss).

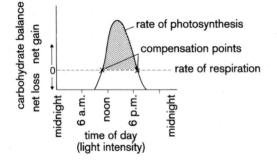

compensation point *The two points at which the rates of photosynthesis and of respiration are equal.*

competition The interaction among organisms of the same species (intraspecific competition) or organisms of different species (interspecific competition) seeking a common resource such as food or light, which is in limited supply in the area occupied by the **community**. Competition often results in the elimination of one organism by another, or even in the elimination of one species as happens when two species of *Paramecium* compete for food.

compound A chemical formed by the combination of **elements** with the component **atoms** occurring in fixed proportions. The basic unit of a compound is the **molecule,** whose formation requires a chemical reaction. By contrast, mixtures have variable proportions of component atoms and can be separated by physical means.

conception An alternative term for **fertilization**.

conditioned reflex A learned **response** to a **stimulus**. It is usually as a result of the repeated association of the stimulus, which may be neutral, to a particular effect that is related to the learned response. For example, a rat may learn to press a lever when hungry as a result of learning to associate the lever's movement with the delivery of food. *See also* **sensitivity**.

cone **1.** A reproductive structure of *gymnosperms*, e.g. pines.
   2. A light-sensitive **neurone** in the **retina** of most vertebrate eyes. Sensitive in bright light, cones can detect colour.

connective tissue Supporting and packing **tissue** in vertebrates, consisting mainly of **collagen** fibres, in which are embedded more complex structures, such as blood vessels, **neurones**, etc.

continuous flow processing An industrial process in **biotechnology**, in hich the **substrates** flow continuously into the **fermentation** vessel and the product flows continuously out. This is made possible by using immobilized **enzymes**, which can be used over and over again (*see* **immobilization**).
   Continuous flow processing is more efficient than **batch processing** since the enzymes can be used repeatedly, the product does not have to be separated from the enzymes, and no time-consuming turn-around is involved.

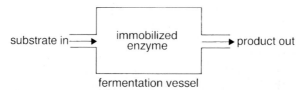

substrate in ⟶ immobilized enzyme ⟶ product out

fermentation vessel

continuous flow processing

continuous variation *See* **variation**.

contractile vacuole A small sac in the **cytoplasm** of freshwater *Protista*, the function of which is **osmoregulation**, i.e., in response to water-entry by **osmosis**, the vacuole expands as it fills with water, and then contracts, discharging its contents out of the cell.

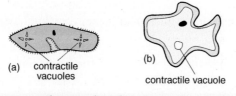

(a)   contractile vacuoles

(b)   contractile vacuole

contractile vacuole *The contractile vacuoles of (a)* Paramecium *and (b)* Amoeba.

control experiment A scientific test in which the factor under investigation is kept constant, so that the result of another test in which this factor is varied can be compared. *See also* **scientific method**.

copulation The coupling of male and female animals for the purpose of fertilization. In humans the **penis** is inserted in to the **vagina** and **spermatazoa** are released.

corm An organ of **vegetative reproduction** in flowering plants consisting of an underground **stem** containing a food store and buds that develop into new plants. Examples of corms include those of the crocus and gladiolus.

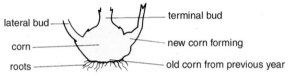

lateral bud
terminal bud
corn
new corn forming
roots
old corn from previous year

corm *Section through a crocus corm.*

cornea Transparent tissue at the front surface of the vertebrate eye, continuous with the **sclerotic** and involved in focusing the image on the **retina**.

cortex 1. (in animals) The outer layer of an organ such as the mammalian kidney. *See also* **medulla**.
2. (in plants) The layer of cells between the **epidermis** and the **vascular bundle**. Cortex cells are packing and supporting tissue, and in some cases, may store food. *See also* **leaf**, **root**, **stem**.

cotyledon An embryonic leaf within a **seed**, which supplies food during **germination**, and in some plants is brought above the soil to carry out photosynthesis for a time before withering. Flowering plants with one cotyledon are called **monocotyledons**, and those with two are called **dicotyledons**.

courtship behaviour A type of animal behaviour that establishes contact between the sexes and thus increases the chance of successful breeding. Examples are bird song and display.

cranial Relating to activities and parts of the body connected with the **brain** and **cranium**.

cranium The bones of the vertebrate skull that enclose and protect the **brain**. *See* **endoskeleton**.

crop rotation The practice of growing a different crop in the same area in successive years in order to prevent **soil depletion**. Since different plants have different **mineral salt** requirements, changing the crop annually prevents depletion of one particular **mineral salt**. Another benefit is that since different plants have different root lengths, they absorb mineral salts from different soil depths. *Leguminous plants* such as peas, beans or clover are often included in rotations because of the **nitrogen fixation** within their **root nodules**. A typical crop rotation might be wheat/turnips/barley/clover/wheat, and so on.

crossing over *See* **meiosis**.

cuticle A noncellular layer secreted by the **epidermis** of above-ground plant structures, and by many invertebrates. Plant cuticles reduce water loss by **transpiration**, while invertebrate cuticles afford protection against mechanical damage and may also retain or repel water.

cytoplasm That part of the **protoplasm** of a cell bounded by the cell membrane (*see* **cell**) but excluding the **nucleus**.

D

deamination Removal of the amino ($-NH_2$) group from surplus **amino acids**. In mammals, this occurs in the liver, the amino group automatically changing to the toxic compound ammonia (NH_3), which is then converted to **urea** and excreted. The remaining carbon-containing group is converted to useful **carbohydrate**.

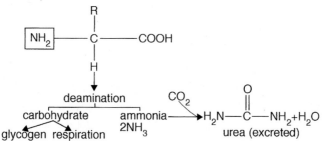

deamination *The results of deamination in mammalian liver. (R is a variable group, depending on amino acid.)*

death rate (of a **population**) The number of deaths, measured in the human population as the number of deaths in one year per 1000 of population. *See also* **human population curve**.

decay The process by which **decomposers** use the organic matter of dead organisms as a source of energy. *See also* **carbon cycle, nitrogen cycle**.

deciduous teeth *See* **milk teeth**.

decomposers **Heterotrophic** organisms that cause the breakdown of dead animals and plants, and by so doing release their constituent compounds, which can be used by other organisms. Soil decomposers include bacteria, earthworms, etc. *See also* **carbon cycle, nitrogen cycle**.

denaturation Changes occurring in the structure and functioning of **proteins** (including **enzymes**) when subjected to extremes of temperature or **pH**.

dendrite or dendron *See* **neurones, synapse**.

denitrification The conversion, by soil *denitrifying bacteria*, of nitrates into nitrogen, which can re-enter the atmosphere. *See also* **nitrogen cycle**.

dental formula A formula describing the **dentition** of a mammal. It is expressed by writing the number of **teeth** in the upper jaw of one side of the mouth over the number of teeth in the lower jaw on one side. The dental formula relates to an adult mammal with the full number of teeth. The total number of teeth is found by doubling the dental formula. *See also* **carnivore**, **herbivore**.

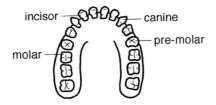

dental formula *Teeth in human upper jaw. Dental formula: incisor 2/2, canine 1/1, premolar 2/2, molar 3/3, total number of teeth = 2 × dental formula = 2 × 16 = 32.*

dentition The numbers and types of **teeth** in a mammal, described by a **dental formula**. Dentition reflects an animal's diet, i.e. an animal has the type of teeth best suited to deal with the type of food on which it feeds. *See also* **carnivore**, **herbivore**, **omnivore**.

deoxygenated blood *See* **vein**.

deoxyribonucleic acid *See* **DNA**.

dialysis *See* **kidney machine**.

diaphragm A dome-shaped muscle separating the **thorax** and **abdomen** in mammals. Contraction and relaxation of the diaphragm is important in lung ventilation. *See also* **breathing**.

diastema A hard toothless gap in the mouth of many **herbivores**, allowing the tongue to manipulate food more easily.

diastole *See* **heartbeat**.

dicotyledons One of the two subsets of flowering plants, the other being **monocotyledons**. The characteristics of dicotyledons are:
(a) two **cotyledons** in the seed;
(b) network of branching **veins** in the leaves;
(c) broad leaves;
(d) ring of **vascular bundles** in the stem;
(e) flower parts in fours or fives or multiples of these numbers.
Examples of dicotyledons include hardwood trees, fruit trees, herbaceous plants.

diffusion The movement of particles from a region of high concentration to a region of lower concentration until an even distribution is reached. Diffusion occurs when two different particle concentrations are adjacent. The difference in concentration that causes diffusion is called a *concentration gradient*. The greater the concentration gradient, the greater is the rate of diffusion. If no concentration gradient exists, diffusion does not occur, and the situation is described as *equilibrium*.

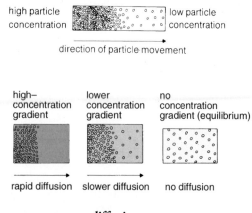

diffusion

Diffusion is the method by which many substances enter and leave living organisms, and are transported within and between cells. Examples are (a) uptake of water by plants from soil, (b) **gas exchange** between plants and the atmosphere, and (c) gas exchange between blood and respiring cells.

Where diffusion is too slow for a particular function, substances can be transported more rapidly by **active transport**. *For a special case of diffusion see* **osmosis**.

digestion The breakdown by **enzyme** action of large insoluble food particles into small soluble particles, prior to **absorption** and **assimilation**. In many animals, including mammals, digestion and absorption occur in the **alimentary canal**. *See table opposite.*

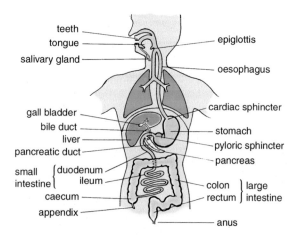

digestion The human alimentary canal.

Location	Glands	Enzyme	Substrate	Product
Mouth	Salivary	Amylase	Starch	Maltose
Stomach	Gastric	Pepsin	Protein	Peptides
		Rennin	Milk protein	Coagulated milk
Duodenum	Pancreas	Amylase	Starch	Maltose
		Lipase	Fats	Fatty acids + glycerol
		Trypsin	Protein + peptides	Amino acids
Ileum		Lactase	Lactose	Glucose + galactose
		Lipase	Fats	Fatty acids + glycerol
		Maltase	Maltose	Glucose + fructose
		Peptidase	Peptides	Amino acids

digestion *Digestive enzymes in humans.*

diploid (used of a nucleus, cell or organism) Having the full complement of **chromosomes**, these occurring in pairs of **homologous chromosomes**. All animal cells, except **gametes**, are diploid. Gametes contain half the diploid number (**haploid**) as the result of **meiosis**. *See also* **fertilization**, **mitosis**.

disaccharides Double **sugar** carbohydrates consisting of two **monosaccharides** linked together by bonds. For example, **maltose** is two **glucose** units joined together, while *sucrose* is one glucose unit linked with one fructose unit.

disaccharides *Chemical structure.*

disinfectant A chemical used to kill **microbes** outside the body, for example on surfaces such as floors.

division A unit used in the **classification** of plants. It is the equivalent of the term **phylum** used in the classification of animals. Divisions and phyla consist of one or more **classes**.

DNA (deoxyribonucleic acid) **Nucleic acid** that is the major constituent of **genes** and hence **chromosomes**. DNA consists of a double polynucleotide chain twisted into a *helix* (spiral) the two chains being held together by bonds between nitrogen *base pairs*. These nitrogen bases can only link as complementary pairs: *thymine* with *adenine* and *guanine* with *cytosine*. The numbers and sequence of base

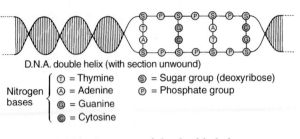

D.N.A. double helix (with section unwound)

Nitrogen bases {
ⓣ = Thymine   ⓢ = Sugar group (deoxyribose)
ⓐ = Adenine   ⓟ = Phosphate group
ⓖ = Guanine
ⓒ = Cytosine
}

DNA *Structure of the double helix.*

pairs in the DNA polynucleotide chain represent coded information (the **genetic code**), which acts as a blueprint for the transfer of hereditary information from generation to generation.

DNA replication The formation of two new molecules of **DNA** during **mitosis**, each of which has exactly the same sequence of bases as the parent molecule. The process requires the four *nucleotides* (*see* **nucleic acids**), the appropriate **enzymes** and **ATP** for **energy**.

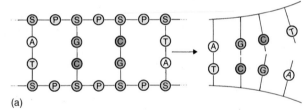

(a)

DNA *replication (a) The DNA double helix unwinds and unzips.*

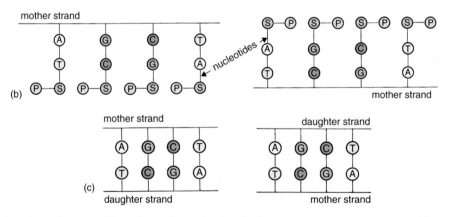

DNA replication (b) *Each single chain then links with the appropriate nucleotide raw materials, which are available in the cell. (c) After attachment of the bases, enzymes link the adjacent S and P molecules into two DNA double chains, which are identical to each other and to the original 'mother' chain.*

dominant (used of one of a pair of **alleles**) Always expressed in a **phenotype**, the other allele being described as **recessive**. *See also* **backcross, codominance, incomplete dominance, monohybrid inheritance**.

dorsal Relating to features of, on, or near, that surface of an organism that is normally directed upwards, although in humans it is directed backwards. *Compare* **ventral**.

Down's syndrome **or** trisomy 21 A human abnormality caused by a **mutation** in which the **ovum** has an extra **chromosome** on chromosome 21. Thus the resulting child has 47 chromosomes in each cells instead of the normal 46. This results in physical and mental retardation. *See also* **amniocentesis**.

drug abuse The habitual consumption of substances that affect the **nervous system**. Such drugs include LSD, heroin and solvents (glue-sniffing). Overdosing on drugs can be fatal.

duodenum The first part of the mammalian **small intestine** leading from the **stomach** via the pyloric **sphincter**. The duodenum receives *pancreatic juice* from the **pancreas** and **bile** from the **liver**, and is an important digestive site.

  The pancreatic juice contains enzymes that continue the **digestion** of food arriving from the stomach. Bile contains *bile salts*, which emulsify fat, forming small fat droplets, thus increasing the surface area available for **lipase** action.

$$\text{Starch} \xrightarrow{\text{Amylase}} \text{Maltose}$$

$$\text{Protein} \xrightarrow{\text{Trypsin}} \text{Peptides} \longrightarrow \text{Amino acids}$$

$$\text{Fat} \xrightarrow{\text{Lipase}} \text{Fatty acids} + \text{Glycerol}$$

From the duodenum, the semidigested food is forced by **peristalsis** into the **ileum**.

E

ear The organ of hearing and balance in vertebrates. Hearing is a sensation produced by vibrations or sound waves, which are converted into **nerve impulses** by the ear and transmitted to the brain.

Outer ear: the **pinna** is a funnel-shaped structure that directs sound waves into the ear and along the **auditory canal**, at the end of which is a very thin membrane, the *eardrum* (**tympanum**), which is made to vibrate by the sound waves.

Middle ear: this is an air-filled cavity connected to the back of the mouth (**pharynx**) by the **eustachian tube**, an arrangement that allows air into the middle ear ensuring equal air pressure on both sides of the eardrum.

Within the middle ear, there are three tiny bones, the **ossicles**, named after their shapes: *malleus* (hammer), *incus* (anvil), and *stapes* (stirrup).

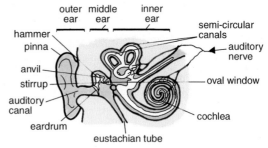

ear *Structure of the human ear.*

The vibrations of the eardrum are transmitted through and amplified by the ossicles, the stapes finally vibrating against a membrane called the **oval window**, which separates the middle and inner ears.

Inner ear: this is fluid-filled and consists of the **cochlea** and **semicircular canals**.

The vibration of the stapes against the **oval window** sets up waves in the fluid of the cochlea. These waves stimulate **receptor** cells (*hair cells*) causing nerve impulses to be sent via the **auditory nerve** to the brain, where they are interpreted as sounds.

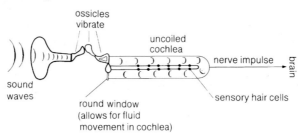

ear *Sound waves are converted into nerve impulses.*

Balance is maintained by the semicircular canals in association with information received from the eyes and muscles. The semicircular canals contain fluid and receptor cells, which are stimulated by movements of the fluid during changes in posture. The nerve impulses initiated by these cells travel to the brain along the auditory nerve and trigger **responses**, which cause the body to maintain normal posture.

ecdysis *See* **exoskeleton.**

ecology The study of the ways in which **communities** of plants and animals interact with one another and with their nonliving environment. *See also* **ecosystem.**

ecosystem A **community** of organisms interacting with each other and with their nonliving **environment**, i.e. a natural unit consisting of living parts (plants and animals) and nonliving parts (light, water, air, etc.).

Habitat + Community → Ecosystem

Ecosystems can be lakes, ocean, forests, etc. The driving force behind all ecosystems is the flow of energy originating from the Sun.

ecosystem management The use of good management practices in order to conserve the **environment**, for example, by taking the appropriate measures to reduce pollution, controlling the use of fertilizers and pesticides and adopting sensible methods of agriculture such as **crop rotation.**

ectoparasite A **parasite** living on the exterior of another organism, its *host.* Lice are examples of ectoparasites. *See also* **endoparasite.**

ectotherm An animal whose body temperature varies with the temperature of its environment. All animals, excluding birds and mammals, are ectotherms, although they are often called *cold blooded. Compare* **endotherm.**

effector A specialized animal tissue or organ that performs a **response** to a **stimulus** from the environment. Examples are muscles and **endocrine glands**. *See also* **sensitivity.**

electron The negatively charged elementary particle that occurs in all **atoms.**

element A pure chemical that cannot be broken down into simpler substances. An element is composed of **atoms** all of which have identical chemical properties and behaviour. There are 92 naturally occurring elements.

embryo **1.** A young animal developed from a **zygote** as a result of repeated **cell division**. In mammals, the embryo develops within the female **uterus**, and in the later stages of pregnancy is called a **foetus**.
   2. A young flowering plant developed from a fertilized **ovule**, which in seed plants is enclosed within a **seed**, prior to **germination**.

embryo sac The structure within the **ovules** of flowering plants in which the female **gametes** are located. *See also* **carpel**.

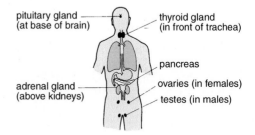

endocrine glands *The main endocrine glands in the human body.*

endocrine glands or ductless glands Structures that release chemicals called **hormones** directly into the bloodstream in vertebrates and some invertebrates. The rate of secretion of hormones is often a response to changes in internal body conditions but may also be a response to changes in the **environment**. *Compare* **exocrine glands**.

endoparasite A **parasite** living inside the body of another organism, its *host*. An example is the tapeworm. *See also* **ectoparasite**.

endoskeleton or internal skeleton The **skeleton** lying within an animal's body, for example, the bony skeleton of vertebrates. Endoskeletons provide shape, support, and protection and work together with **muscles** to produce movement. *See* figure opposite. *Compare* **exoskeleton**.

endotherm An animal that can maintain a constant narrow range of body temperature despite fluctuations in the temperature of the environment. Mammals and birds are endotherms, although they are often called warm blooded. *Compare* **ectotherm**; *see also* **temperature regulation**.

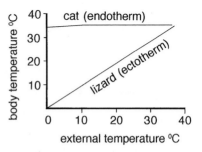

endotherm *The relationship between external temperature and body temperature in an endotherm and an ectotherm.*

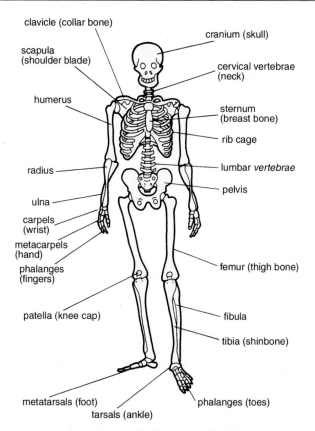

clavicle (collar bone)

cranium (skull)

scapula
(shoulder blade)

cervical vertebrae
(neck)

humerus

sternum
(breast bone)

rib cage

radius

lumbar *vertebrae*

pelvis

ulna

carpels
(wrist)

metacarpels
(hand)

phalanges
(fingers)

femur (thigh bone)

patella (knee cap)

fibula

tibia (shinbone)

metatarsals (foot)

phalanges (toes)

tarsals (ankle)

endoskeleton *The human endoskeleton.*

energy The ability to do work. In living organisms that work pertains to the seven characteristics of life: movement, feeding, **reproduction, excretion, growth, sensitivity, respiration.**

The types of energy are: heat, light, sound, electrical, chemical, nuclear, potential (stored) and kinetic (moving). Energy can be neither created nor destroyed, but it can be changed from one form into another. This scientific law is called the law of conservation of energy. Examples include:

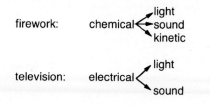

firework: chemical → light, sound, kinetic

television: electrical → light, sound

This concept of energy interconversion is important to living organisms, since green plants convert the light energy of sunlight into the chemical/potential energy of food, via the reaction of **photosynthesis**.

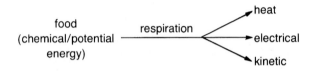

Other organisms can then release that chemical/potential energy via the reaction of **respiration** and convert it into other useful forms. For example:

food
(chemical/potential
energy) ———— respiration ————<
heat
electrical
kinetic

environment The surroundings in which organisms live and which influence the distribution and success of organisms. Many factors contribute to the environment, including (a) nonliving **abiotic factors**, e.g. temperature, light, **pH**, etc., and (b) living **biotic factors**, e.g. **predators**, **competition**. The interaction of these factors determines the conditions within **habitats**, and 'selects' the **communities** of organisms that are best suited to these conditions.

enzymes **Proteins** that act as biological **catalysts** within cells. Catalysts are substances that accelerate chemical reactions. In cells there may be hundreds of reactions occurring, each one requiring a particular enzyme.

A + B ——×—▶ no reaction

enzyme
A + B ————▶ C + D
reactants   products

Enzymes catalyse either *synthesis* by which complex compounds are formed from simple molecules, or *degradation* by which complex molecules are broken down to simple subunits by **hydrolysis**. *See also* **digestion**.

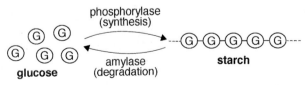

enzymes *The synthesis and degradation of starch.*

Enzyme characteristics:

(a) Enzymes are proteins.

(b) Enzymes work most efficiently within a narrow temperature range. Thus human enzymes work best at 37 °C (body temperature) and this is called the optimum temperature. Above or below this temperature their efficiency decreases, and at temperatures above 45 °C most enzymes are destroyed (**denaturation**).

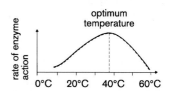

enzymes *Graph showing the effect of temperature upon the rate of enzyme activity.*

(c) Enzymes have a **pH** at which they work most efficiently; this is known as the optimum pH. For example, the saliva enzyme salivary amylase works best at neutral or slightly acid pH. The stomach enzyme **pepsin** will only function in an acidic pH, while the enzyme **trypsin** in the intestine favours an alkaline pH.

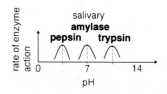

enzymes *Graph showing the effect of pH upon enzyme activity.*

(d) The rate of an enzyme-catalysed reaction increases as the enzyme concentration increases.

enzymes *Graph showing the effect of enzyme concentration upon the rate of reaction.*

(e) The rate of an enzyme-catalysed reaction increases as the **substrate** concentration increases, up to a maximum point.

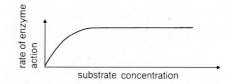

enzymes *Graph showing the effect of enzyme concentration upon the rate of reaction.*

(f) Normally an enzyme will catalyse only one particular reaction, a property called *specificity*. For example, the enzyme catalase can only degrade the compound hydrogen peroxide.

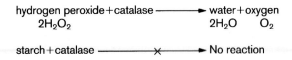

Most enzymes are named by adding the suffix -*ase* to the name of the enzyme's substrate (or by changing -*ose* to -*ase*). For example, maltase acts on maltose; urease acts on **urea**, etc.

enzyme mechanism The way in which an enzyme and its substrate combine. Enzyme action is explained by the *lock and key hypothesis* in which the enzyme is thought of as a lock into which only certain keys (the substrate molecules) can fit. In this way, the enzyme and substrate are brought together and the reaction can occur.

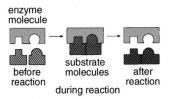

enzyme mechanism *Sequence illustrating an enzyme-catalysed synthesis. Reversing the order of the diagrams shows how an enzyme-catalysed degradation occurs.*

epidermis The protective outermost layer of cells in an animal or plant. The epidermis of many **multicellular** invertebrates is one cell thick and is often covered with a **cuticle**. In most vertebrates the epidermis is the outer layer of skin. Land vertebrates may have several layers of dead cells. In plants, the epidermis is one cell thick. Aerial (above-ground) structures may have a cuticle. *See also* **leaf**, **root**, **stem**.

epiglottis The flap of cartilage and membrane at the base of the tongue on the ventral wall of the **pharynx**. It closes the **trachea** during swallowing. *See* **digestion**.

epithelium Lining tissue in vertebrates consisting of closely packed layers of cells, covering internal and external surfaces. Examples are the skin and the lining of the breathing, digestive and urinogenital organs. Epithelia may also contain specialized structures, e.g. **cilia**, **goblet cells**.

erythrocyte *See* **red blood cell**.

eustachian tube The tube connecting the **middle ear** to the **pharynx** in **tetrapods**. It is important for equalizing air pressure on either side of the **eardrum**. *See* **ear**.

evolution The development of complex organisms from simpler ancestors, occurring over successive generations. *See also* **natural selection**.

excretion Elimination of the waste products of **metabolism** by living organisms. The main excretory products are water, carbon dioxide and nitrogenous compounds, e.g. **urea**. In simple organisms excretion occurs through the cell membrane (*see* **cell**) or **epidermis**; in higher plants via the leaves; while most animals have specialized excretory organs. For example, in man the lungs excrete water and carbon dioxide, and the kidneys excrete urea.

exercise Any physical activity or bodily exertion that contributes to good health. Evidence suggests that regular exercise reduces the risk of heart disease, probably as a result of improved blood circulation. *See also* **circulatory system**.

exocrine glands Structures in vertebrates that release secretions to **epithelial** surfaces via **ducts**. Examples are sweat glands; salivary glands. *Compare* **endocrine glands**.

exoskeleton or external skeleton A **skeleton** lying outside the body of some invertebrates, for example, the **cuticle** of insects and the shells of crabs. Some organisms shed and renew their exoskeletons periodically to allow growth, a process known as *moulting* or **ecdysis**. *Compare* **endoskeleton**.

(a)   (b)

exoskeleton *The exoskeletons of (a) an insect and (b) a crustacean.*

eye A sense organ that responds to light, ranging from very simple structures in invertebrates to the complex organs of insects and vertebrates. Eye muscles enable the eye to move up and down and from side to side. The **sclerotic** is a tough protective layer, which at the front of the eye forms the transparent **cornea**.

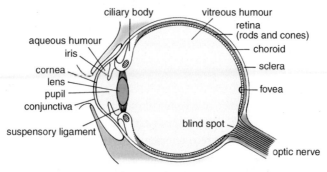

eye *Vertical section through the human eye.*

The **choroid** is a black-pigmented layer under the sclerotic, rich in blood vessels supplying food and oxygen to the eye.

The **retina** is a layer of **neurones** that are sensitive to light. There are two types of cell in the retina, named by their shape:

(a) **rods** are very sensitive to low intensity light and are particularly concentrated in the eyes of nocturnal animals;

(b) **cones** are sensitive to bright light. There are different types that are stimulated by different wavelengths of light and are thus responsible for *colour vision*. Animals whose retinas lack cones are colour blind, while human colour blindness is caused by a defect in the cones.

The **fovea** (*yellow spot*) is a small area of the retina containing only cones in great concentration and giving the greatest degree of detail and colour.

The **blind spot** is that part of the retina at which nerve fibres connected to the rods and cones leave the eye to enter the **optic nerve**, which leads to the brain. Since there are no light-sensitive cells at this point, an image formed at the blind spot is not registered by the brain.

The **lens** is a transparent biconvex structure that can change curvature. It is held in place by **suspensory ligaments** that are attached to the **ciliary muscles**, the contraction or relaxation of which alters the shape of the lens, allowing both near and distant objects to be focused sharply. This is called **accommodation**.

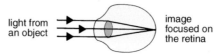

eye *The lens focuses light on the retina.*

The **iris** is the coloured part of the eye, containing muscles that vary the size of the *pupil*, the hole through which light enters the eye. In poor light, the pupils are wide open (dilated), to increase the brightness of the image.

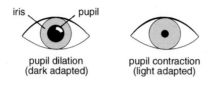

pupil dilation
(dark adapted)

pupil contraction
(light adapted)

eye *The pupil dilates and contracts according to the intensity of the light.*

In bright light the pupils are contracted to protect the retina from possible damage. This mechanism is an example of a **reflex action**.

Because the pupil is small, light rays enter the eye in such a way that the image projected at the retina is upside down (inverted). This inversion of the image is corrected by the brain.

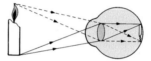

eye *The image is smaller than the object, and is inverted.*

The **aqueous humour** and **vitreous humour** are fluids that fill the chambers of the eye. They help to maintain shape, focus light, and allow nutrients, oxygen and wastes to diffuse to and from the eye cells.

F

F$_1$ generation (first filial generation) The first generation of **progeny** obtained in breeding experiments. Successive generations are called F$_2$, etc. *See also* **monohybrid inheritance**.

faeces The solid or semisolid remains of undigested food, bacteria, etc, which are formed in the **colon** of vertebrates and expelled via the **anus**.

family The unit used in the **classification** of living organisms consisting of one or more genera (singular **genus**).

fats or lipids **Organic compounds** containing the elements carbon, hydrogen, oxygen. Fats consist of three *fatty acid* molecules (which may be the same or different) bonded to one *glycerol* molecule.

Fat deposits under the skin act as long term energy stores, yielding 39 kJ/g when respired; these deposits also provide heat insulation.

Fat is an important constituent of the cell membrane (*see* **selectively permeable membrane**) and its insolubility in water is utilized in the waterproofing systems of many organisms.

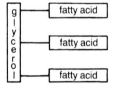

fats *Structure of a fat molecule.*

femur **1.** The part of an insect limb nearest to the body.
2. The thighbone of **tetrapod** vertebrates. *See also* **endoskeleton**.

fermentation The degradation of **organic compounds** in the absence of oxygen for the purpose of energy production, by certain organisms, particularly bacteria and yeasts. Fermentation is a form of anaerobic **respiration**. *See also* **brewing**.

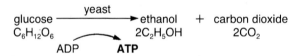

fermentation *The fermentation of sugars by yeast produces the by-products ethanol and carbon dioxide.*

fertilization The fusing of **haploid gametes** during **sexual reproduction** resulting in a single cell, the **zygote**, containing the **diploid** number of chromosomes. It occurs in both animals and plants. In animals, fertilization is either external or internal.

External fertilization occurs when the gametes are passed out of the parents and fertilization and development take place independently of the parents. External fertilization is common in aquatic organisms where the movement of water helps the gametes to meet.

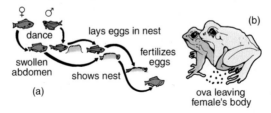

fertilization *External fertilization in (a) fish: the stickleback; and (b) amphibians: the toad.*

Internal fertilization is particularly associated with terrestrial animals, for example, insects, birds and mammals, and involves the union of the gametes within the female's body. The advantages of internal fertilization are (a) the **sperms** are not exposed to unfavourable dry conditions, (b) the chances of fertilization occurring are increased, and (c) the fertilized **ovum** is protected within a shell (as in birds) or within the female body (as in mammals).

fertilization *Internal fertilization in an insect: the locust.*

Sperm cells, produced in the **testes**, are passed out of the **penis** during **copulation** in which the penis is inserted in the **vagina**. The sperms move through the **uterus** and, if an ovum is present in an **oviduct**, fertilization can occur there.

The fertilized ovum (zygote) continues moving towards the uterus, dividing repeatedly as it does so. On arrival at the

fertilization *The human reproductive organs.*

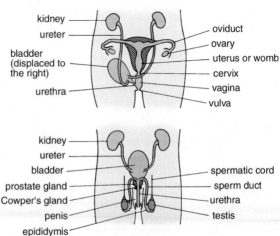

uterus, the zygote, by now a ball of cells, becomes embedded in the prepared wall of the uterus. This is called **implantation**. Further development of the **embryo** occurs in the uterus. *See* **pregnancy, birth.**

Fertilization in plants. In flowering plants, after **pollination, pollen** grains deposited on **stigmas** absorb nutrients. *Pollen tubes* then grow down through a narrow region called the *style* and enter the **ovules** through the **micropyles.** The tip of each pollen tube breaks down and the male **gamete** enters the ovule and fuses with the female gamete. After fertilization, the ovule, containing the plant embryo, develops into a **seed**, and the **ovary** develops into a **fruit.** *See* **flower.**

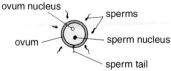

fertilization
Fertilization of the ovum.

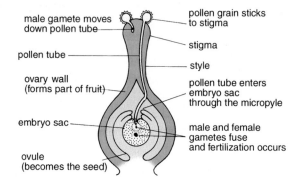

fertilization *Fertilization in plants.*

fertilizer Any substance added to soil to increase the quantity or quality of plant growth. When crops are harvested, the natural circulation of soil **mineral salts** is disturbed, i.e., mineral salts absorbed by plants are not returned to soil. This is called **soil depletion** and may render the soil infertile. Fertilizers replenish the soil and are of two types:
(a) organic fertilizers such as compost, sewage;
(b) inorganic fertilizers such as ammonium sulphate.

fibrinogen A soluble **plasma protein** involved in **blood clotting**.

fibula The posterior of two bones in the lower hind-limb of **tetrapods**. In humans it is the outer bone of the leg below the knee. *See* **endoskeleton.**

fitness The state of being in healthy physical condition. Fitness is achieved and maintained by adopting a healthy lifestyle, for example by having a **balanced diet**, taking regular exercise, and avoiding alcohol, smoking and drug abuse.

flagellum A microscopic motile thread projecting from certain cell surfaces and causing movement by lashing back and forth. Flagella are usually larger than **cilia** and less numerous, and are responsible for locomotion in many **unicellular** organisms and reproductive cells.

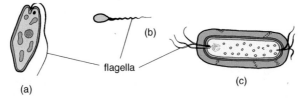

flagellum *Flagella in (a) Euglena, (b) motile sperm and (c) motile bacterium.*

flower The organ of **sexual reproduction** in flowering plants (angiosperms).
   Insect-pollinated flowers have brightly coloured and scented petals, and usually have a *nectary* at the base of the flower. The nectary contains a sugar solution called *nectar*. The **stamens** and **carpels** (with sticky **stigmas**) are within the flower. These adaptations favour insect pollination.
   Wind-pollinated flowers are small, often green and unscented, and do not have nectaries. The **anthers** and feathery stigmas dangle out of the flowers when ripe thus facilitating wind pollination. *See also* **fertilization, pollen, pollination**.

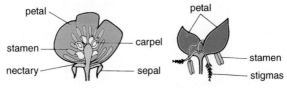

flower *The structure of (a) an insect-pollinated flower and (b) a wind-pollinated flower.*

foetus The mammalian **embryo** after development of the main features. In humans this is after about three months of **pregnancy**.

follicle-stimulating hormone (FSH) A **hormone** secreted by the vertebrate **pituitary gland**. *See* **ovulation**.

food calorimeter A device for measuring the **energy** content of food.
A weighed food sample is ignited using a heating filament. In the
presence of oxygen, the food sample is completely combusted, the released
energy being transferred to the surrounding water, which shows a rise
in temperature.

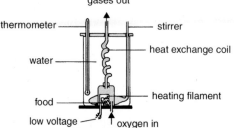

food calorimeter

From this rise in temperature, it is possible to calculate the energy
content of the food in **kilojoules** per gram (kJ/g).

food	energy content
protein	17 kJ/g
carbohydrate	17 kJ/g
fat	39 kJ/g

food calorimeter *The energy content of the different kinds of food.*

food capture The method by which organisms obtain food. Many
heterotrophic organisms have developed very specialized methods and
structures for obtaining food; a variety of examples is given below.
(a) *Mammals without teeth.* Anteaters have a long and sticky tongue for
    catching ants. Blue whales have modified mouth parts to filter **plankton**
    out of water.
(b) *Filter feeding.* Like the blue whale, many aquatic organisms, especially
    invertebrates, filter plankton.

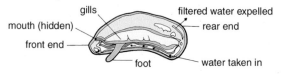

food capture *Filter feeding: Mytilus (the edible mussel)*
shown with one shell removed.

(c) *Feeding by sucking.* Houseflies pass **saliva** out onto their food, e.g.
    sugar. **Digestion** begins immediately and the resulting liquid is then

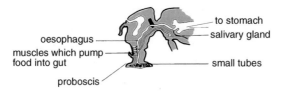

food capture *The housefly sucks up dissolved food.*

taken in by a sucking pad on an elongated structure called a *proboscis*.

Female mosquitoes pierce the skin, injecting a fluid that prevents **blood clotting**, and then suck up blood. This feeding mechanism can cause the disease malaria, since the **parasite** involved may be transmitted during feeding.

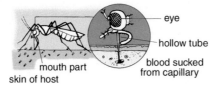

food capture *The female mosquito sucks blood.*

Butterflies feed on the nectar produced by flowering plants. They suck the nectar from the flower by means of a long tube-like proboscis, which remains coiled when not in use.

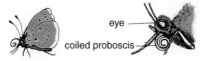

food capture *The buttefly's long, coiled proboscis.*

Greenflies use a long piercing proboscis to suck plant juices from leaves and **stems**.

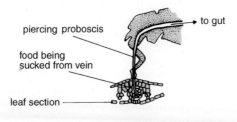

food capture *The greenfly's proboscis.*

(d) *Biting without teeth*. Locusts eat their own weight of plant material every day. They have powerful biting jaws called *mandibles*, which have very hard biting edges that are brought together during feeding in a precise and efficient shearing action. *See also* **carnivore, dentition, herbivore, omnivore**.

mandibles

food capture
The locust's biting jaws.

food chain A food relationship in which **energy** and carbon compounds obtained by green plants via photosynthesis are passed to other living organisms, i.e., plants are eaten by animals, which in turn are eaten by other animals and so on.

green ——▶ **herbivore** ——▶ small ——▶ large
plant   (primary   **carnivore**   carnivore
(producer)   consumer)   (secondary   (tertiary
                       consumer)   consumer)

The arrows indicate 'is eaten by' and the direction of energy flow. An example of such a food chain is:

plants → insects → lizards → snakes

Not all food chains are as long as the above, for example:

grass → sheep → man

grass → antelope → lion

Such simple food chains seldom exist independently; more often several food chains are linked in a more complicated relationship called a *food web*.

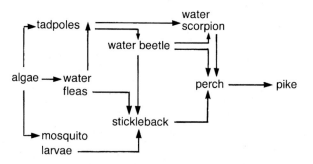

food chain *Part of the food web in a freshwater pond.*

All food webs are delicately balanced. Should one link in the web be destroyed, all the other organisms will be affected. For example, in the pond food web, if the perch disappeared as the result of disease, the pike population would decrease while the water scorpions would increase.

food consumers **Heterotrophic** organisms which, after green plants, occupy the subsequent links in a **food chain**.

food producers **Autotrophic** organisms, mainly green plants, which occupy the first level in a **food chain**.

food tests Chemical tests used to identify the components of a food sample. Some common food tests are shown below.

protein + Biuret reagent → violet/purple colour
(blue)

heat
reducing + Benedict's reagent → green/yellow/brick
sugar (blue) red colour
starch + iodine solution → blue/black colour
(brown)
fat + water + ethanol → white emulsion
(clear)
vitamin C + dichlorophenolindo- → DCPIP
phenol (DCPIP) (clear)
(blue)

food web *See* **food chain**.

fossil The mineralized remains of animal and plant tissue from previous geological ages, which have been preserved in the Earth's crust. The study of fossils is called **palaeontology**, and has supplied important evidence for **evolution**.

fossil fuels The fossilized remains of plants and animals which, over millions of years, have been transformed into oil, coal and natural gas. *See also* **carbon cycle**, **greenhouse effect**, **acid rain**.

fovea or yellow spot The area of the **retina** in some vertebrate **eyes**, specialized for acute vision. It contains numerous **cones** but no **rods**.

fruit The ripened **ovary** of a **flower**, enclosing seeds, formed as the result of **pollination** and **fertilization**. The fruit protects the seed and aids its dispersal.

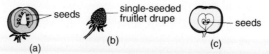

fruit *(a) tomato (b) blackberry (c) apple.*

fruit and seed dispersal The methods by which most flowering plants spread seeds far away from the parent plant, thus (a) avoiding **competition** for resources and (b) ensuring wide colonization, so that suitable **habitats** are likely to be encountered by a proportion of seeds. The methods are:

(a) *Wind dispersal*: Air currents carry the fruits or seeds, which usually show an adaptation to increase their surface area.

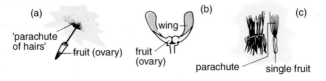

fruit and seed dispersal *Methods of wind dispersal (a) dandelion (b) sycamore (c) groundsel.*

(b) *Animal dispersal*: Hooked fruits, e.g. burdock, stick to animals' coats and may be brushed off some distance from the parent plant. Succulent berries, such as the strawberry, are eaten by animals, and the small hard fruits containing seeds pass through the gut unharmed before being released in the **faeces**.

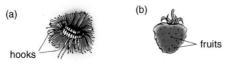

fruit and seed dispersal *Examples of fruit dispersed by animals (a) burdock (b) strawberry.*

(c) *Explosive dispersal*: Unequal drying of part of a fruit causes the fruit to burst, shooting the seeds away from the parent plant.

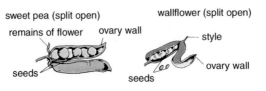

fruit and seed dispersal *Sweet pea and wallflower disperse their seeds by explosion of the fruit.*

FSH *See* **follicle-stimulating hormone**.

fungi A **heterotrophic** plant group that includes the microscopic moulds and **yeasts** such as *Mucor* and *Penicillium* and also the **multicellular** mushrooms and toadstools. Some fungi are **parasites** causing plant disease such as potato blight and Dutch elm disease. In humans, fungi cause athletes foot, ringworm and lung infections. Other fungi are useful to man, for example as a source of antibiotics and in brewing.

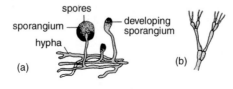

fungi *(a) Mucor (b) Penecillium.*

G

gall bladder A small sac in or near the vertebrate **liver**, in which **bile** is stored. When food enters the **intestine**, the gall bladder contracts and empties bile into the **duodenum** via the **bile duct**. *See also* **digestion**.

gamete A reproductive cell whose nucleus is formed by **meiosis** and contains half the normal **chromosome** number (i.e. they are **haploid** cells). Human male gametes are **spermatozoa** and human female gametes are **ova** (egg cells). Male and female gametes fuse during **fertilization** forming a **zygote** in which the normal chromosome number (**diploid**) is restored.

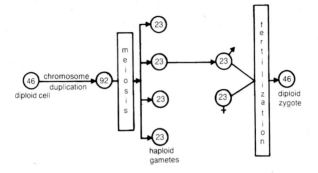

gamete *In humans, the haploid gametes have 23 chromosomes and the diploid zygote, formed after fertilization, has 46.*

gas exchange The process by which organisms exchange gases with the environment for the purpose of **metabolism**. Most organisms require a continuous supply of the gas oxygen for **respiration**:

glucose+oxygen→**energy**+carbon+water
dioxide

In addition, green plants require carbon dioxide for **photosynthesis**:

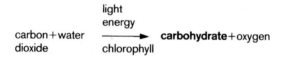

Both reactions use and produce gases that are interchanged between the atmosphere (in the case of land organisms) or water (aquatic organisms).

Gas exchange takes place across surfaces that have the following characteristics:
(a) a large surface area for maximum gas exchange;
(b) the surface is thin to allow easy **diffusion**;
(c) the surface is moist since gas exchange occurs in solution;
(d) in animals, the surface has a good blood supply, since the gases involved are transported via the blood.

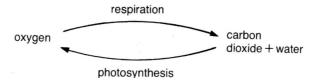

gas exchange *Both respiration and photosynthesis involve gas exhange with the environment.*

Gas exchange in fish occurs across **gills**, which consist of *gill arches* to which are attached numerous gill filaments. Water is taken in via the mouth and passed over the gills where oxygen dissolved in the water is absorbed into blood capillaries while carbon dioxide diffuses into the water.

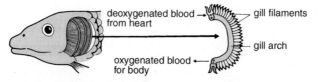

gas exchange *The position and structure of the gills of fish.*

In insects, air enters through pores called **spiracles** and is carried through a branching system of **tracheae** and thus into smaller branches called *tracheoles*, which are in contact with the tissues.

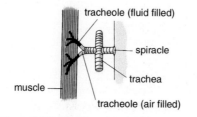

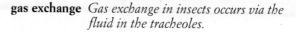

gas exchange *Gas exchange in insects occurs via the fluid in the tracheoles.*

Gas exchange in mammals occurs as the result of *concentration gradients* (*see* **diffusion**) existing between the air in the alveoli and the deoxygenated blood arriving from the heart. These gradients cause diffusion of oxygen from the alveoli into **red blood cells** and diffusion of carbon dioxide from the blood into the alveoli. *See also* **breathing**, **lungs**.

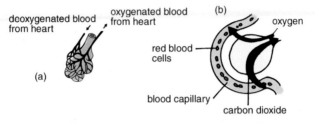

gas exchange *(a) Alveoli and associated blood vessels.*
(b) Gas exchange in the alveolus.

Gas exchange in plants:

(a) Terrestrial plants: gas exchange in leaves and young stems occurs through pores in the **epidermis** called **stomata**. In young roots it occurs by diffusion between the roots and air in the soil. In older stems and roots where bark has formed, gas exchange takes place through gaps in the bark called **lenticels**.

(b) Aquatic plants: submerged plants, e.g. pond-weed, have no stomata, gas exchange occurring by diffusion across the cell membranes (*see* **cell**). Aquatic plants with floating leaves, e.g. water lily, have stomata only on the upper leaf surface.

(c) Nongreen plants: mushrooms, for example, carry out **respiration** but not **photosynthesis**. Gas exchange occurs by diffusion between the plant cells and the surrounding air.

gas exchange *Section throught plant stem showing lenticel.*

gastric Relating to parts and functions of the body connected with the **stomach**.

genes The subunits of **chromosomes**. They consist of lengths of **DNA**, which control the hereditary characteristics of organisms. Genes consist of up to one thousand *base pairs* in a DNA molecule, the particular sequence of which represents coded information. This is known as the **genetic code** and determines the types of proteins synthesized by cells, particularly enzymes, which then dictate the structure and function of cells and tissues, and ultimately organisms. In other words, a cell or an organism is an expression of the genes it has inherited (and the environment in which it lives).

The genetic code is the arrangement of nitrogen base pairs in DNA. Each group of three adjacent base pairs (*triplets*) is responsible for linking together, within the cell, **amino acids** to form protein. The sequence, types and numbers of amino acids determine the nature of the proteins, which in turn determine the characteristics of cells. For example, the base triplet GTA codes for the amino acid histidine while GTT codes for glutamine.

Consider two fruit flies (*Drosophila*), one with a gene X controlling body colour, while the other's body colour is controlled by gene Y.

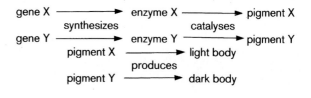

genetic code See **genes**.

genetic engineering The transfer of pieces of **chromosome** from one organism to another. For example, the **gene** for human **insulin** can be inserted into a bacterium, which will then produce insulin. This can be isolated and purified for the treatment of diabetes. *See diagram.*

genetics The study of *heredity*, which is the transmission of characteristics from parents to offspring via the **genes** in the **chromosomes**. Heredity is investigated by performing breeding experiments and then comparing the characteristics of the parents and offspring. Such experiments were first done by Gregor

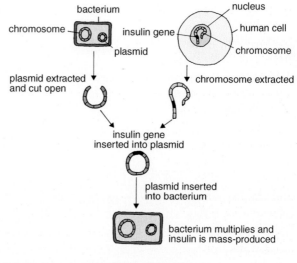

genetic engineering

Mendel in the 1860s using pea plants. *See* **backcross, codominance, incomplete dominance, monohybrid inheritance.**

genotype The genetic composition of an organism, i.e. the particular set of **alleles** in each cell. In breeding experiments, genotypes are represented by symbols, capital letters denoting the **dominant** alleles and small letters denoting the **recessive** alleles. *See also* **monohybrid inheritance.**

genus A unit used in the **classification** of living organisms, consisting of a number of similar **species.**

geotropism A form of **tropism** relative to gravity. Plant shoots grow away from gravity (negative geotropism), but most roots grow downwards and thus show positive geotropism.

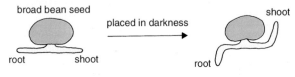

broad bean seed

placed in darkness

shoot

root   shoot

root

geotropism

germination The beginning of growth in **spores** and **seeds**. It normally proceeds only under certain environmental conditions, for example, the availability of water and oxygen, and a favourable temperature. If these conditions are not present, spores and seeds may remain alive for some time without germinating. In this state, known as *dormancy*, the seeds are dormant, i.e. inactive while they await the right conditions.

In flowering plants there are two types of germination: *hypogeal* and *epigeal*. They are distinguished by what happens to the **cotyledons** during development of the seedling. In hypogeal germination, the

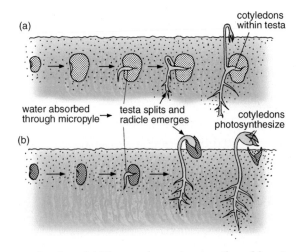

(a)

cotyledons within testa

water absorbed through micropyle

testa splits and radicle emerges

cotyledons photosynthesize

(b)

germination *(a) Hypogeal germination (broad bean). (b) Epigeal germination (French bean).*

cotyledons remain below ground while in epigeal germination the cotyledons are taken above ground. In both cases water is absorbed via the **micropyle**, the **testa** splits open and the **radicle** emerges.

gestation period *See* **pregnancy**.

gills The **gas-exchange** surface of aquatic animals. In fish, gills are usually internal, projecting from the **pharynx**, while in amphibian **larvae**, they are external.

gland A cell or organ that synthesizes chemical substances and secretes them into the body either through a duct or direct to the bloodstream. *See also* **endocrine gland**.

glomerulus (*pl.* **glomeruli**) The knot of blood capillaries within the **Bowman's capsule** of the mammalian **kidney**.

glottis The opening of the **larynx** into the **pharynx** of vertebrates.

glucose A **monosaccharide** carbohydrate, synthesized in green plants during **photosynthesis**, and serving as an important energy source in animal and plant cells. *See also* **monosaccharide**, **respiration**.

glycogen A **polysaccharide** carbohydrate consisting of branched chains of **glucose** units, which is important as an energy store in animals. In vertebrates, glycogen is stored in muscle and liver cells and is readily converted to glucose by enzymes. *See also* **insulin**.

goblet cells Specialized cells in certain **epithelia**, which synthesize and secrete **mucus**. Goblet cells are common in vertebrates, and are found in, for example, the intestinal and respiratory tracts of mammals.

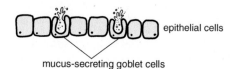

epithelial cells

mucus-secreting goblet cells

goblet cells *Goblet cells in epithelium.*

gonads Animal organs that produce **gametes** and in some cases hormones. Examples are the **ovaries** and **testes**.

Graafian follicle A fluid-filled cavity in the mammalian **ovary** within which the **ovum** develops until **ovulation**.

greenhouse effect The increase in global temperature caused by increasing atmospheric carbon dioxide levels. This has resulted from burning **fossil fuels** and the destruction of large areas of tropical forest (so that less photosynthesis takes place). The increased carbon dioxide concentration traps the radiant energy of the Sun in a similar way to a greenhouse and may result in the polar ice-caps melting and a rise in sea level.

growth The increase in size and complexity of an organism during development from embryo to maturity, resulting from **cell division**, cell enlargement and **cell differentiation**. In plants, growth originates at certain localized areas called *meristems*, while animal growth goes on all over the body.

guard cells Paired cells bordering **stomata** and controlling the opening and closing of the stomata. The **diffusion** of water into guard cells from adjacent **epidermis** cells causes the guard cells to expand and increase their **turgor**. However, they do not expand uniformly, the thicker, inelastic cell walls causing them to bend so that a pair of guard cells draws apart forming a stoma. Diffusion of water from guard cells reverses the process and closes the stoma. Stomata are normally open during the day and closed at night.

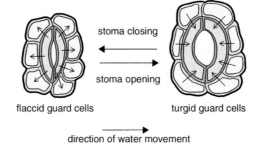

guard cells *Stoma closing and opening.*

gut All or part of the **alimentary canal**.

H

habitat The place where an animal or plant lives, the organism being adapted to the particular conditions within the habitat, which may be a woodland, sea-shore, pond, rockpool, etc. *See* **environment**.

haemoglobin Red pigment containing iron, within vertebrate **red blood cells**. It is responsible for the transport of oxygen throughout the body.

haemolysis The loss of **haemoglobin** from **red blood cells** as a result of damage to the cell membrane (*see* **cell**). This can be caused by several factors including **osmosis**, which can be investigated using human red blood cells that have a **solute** concentration equivalent to a 0.9% sodium chloride solution (*see diagram*).

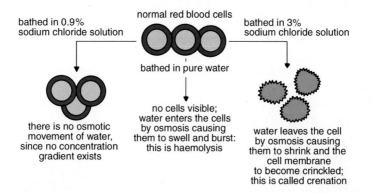

haemolysis *Red blood cells in sodium chloride solution burst or shrink according to the concentration of the solution.*

haemophilia *See* **sex linkage**.

haploid (used of a nucleus, cell or organism) Having a single set of unpaired **chromosomes**. The haploid number is found in plant and animal **gametes** as the result of **meiosis**. *See also* **diploid**.

heart A muscular pumping organ that maintains **blood** circulation, and is usually equipped with **valves** to prevent backward flow. In mammals, the heart has four chambers, consisting of two relatively thin-walled *atria* (or *auricles*), which receive blood, and two thicker-walled *ventricles*, which pump blood out.

The right side of the heart deals only with deoxygenated blood, and the left side only with oxygenated blood. The wall of the left ventricle is thicker and more powerful than that of the right, since it pumps all round the body, while the right ventricle pumps only to the lungs. *See also* **circulatory system, heartbeat.**

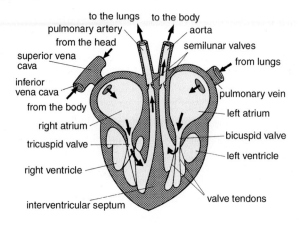

heart *Structure of the mammalian heart.*

heartbeat The alternate contraction and relaxation of the **heart**. In mammals it consists of two phases:

(a) *diastole*: the *atria* (singular, *atrium*) and **ventricles** relax, allowing blood to flow into the ventricles from the atria;

(b) *systole*: the ventricles contract, forcing blood into the **pulmonary artery** and **aorta**. The relaxed atria fill with blood in preparation for the next beat.

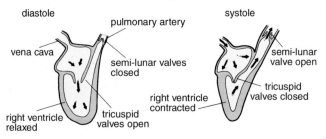

heartbeat *Relaxation and contraction of heart (right side).*

Heartbeat is initiated by a structure in the right atrium called the **pacemaker**, although the rate is controlled by the **medulla oblongata** of the brain, which detects any increase in carbon dioxide in the blood as the result of increased respiration and is also affected by certain hormones, for example, **adrenalin** from the **adrenal gland**. The rate of human heartbeat is measured by counting the **pulse rate**.

hepatic (used of parts of the body) Relating to the **liver** and its functions.

herbivore An animal that feeds on plants. Herbivores, which include sheep, rabbits and cattle, have a **dentition** adapted for chewing vegetation and a gut capable of **cellulose** digestion.

   Most herbivore teeth are grinders; **canines** are usually absent. In herbivores without upper **incisors**, a horny pad of gum combines with the lower incisors in biting vegetation. In most herbivores the lower jaw can move sideways, or backwards and forwards, thus producing the grinding action of the teeth.

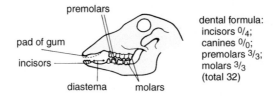

premolars

dental formula:
incisors $0/4$;
canines $0/0$;
premolars $3/3$;
molars $3/3$
(total 32)

pad of gum

incisors

diastema   molars

herbivore *The skull and teeth of a sheep.*

heterotrophic (used of organisms) Able to obtain **organic compounds** (food) by feeding on other organisms. Heterotrophs include all animals and fungi, most bacteria and a few flowering plants. Heterotrophs are also called **food consumers** and can be classified into **carnivores**, **herbivores**, **omnivores**, **saprophytes** and **parasites**. *Compare* **autotrophic**.

heterozygous Having two different **alleles** of the same **gene**. *See* **monohybrid inheritance**.

homeostasis The maintenance of constant conditions within an organism. Examples are the control of blood **glucose** levels by **insulin**, the control of blood water content by **ADH**, and the control of body temperature by the skin, etc.

homologous chromosomes Pairs of **chromosomes** that come together during **meiosis**. They carry **genes** that govern the same characteristics. Homologous pairs of chromosomes are found in all **diploid** organisms, one of the pair coming from the male **gamete** and the other from the female gamete, the pair being united at fertilization.

homozygous or **pure** Having two identical **alleles** for any one **gene**. *See* **monohybrid inheritance**.

hormones **1.** (animal) Chemicals secreted by the **endocrine glands** of animals and transported via the bloodstream to certain organs (*target organs*) where they cause specific effects that are vital in regulating and coordinating body activities. Hormone action is usually slower than nervous stimulation. The table below summarizes the properties of some important human hormones; there are many others.
2. (plant) *See* **plant growth substances**.

Endocrine gland	Hormone	Effects
Pituitary gland	ADH (anti-diuretic hormone)	Controls water reabsorption by the kidneys.
	TSH (thyroid stimulating hormone	Stimulates thyroxine production in the thyroid gland.
	FSH (follicle-stimulating hormone	Causes ova to mature and the ovaries to produce oestrogen.
	LH (luteinizing hormone)	Initiates ovulation and causes the ovaries to release progesterone.
	Growth hormone	Stimulates growth in young animals. In humans, deficiency causes dwarfism and excess causes gigantism.
Thyroid gland	Thyroxin	Controls rate of growth and development in young animals. In human infants, deficiency causes cretinism. Controls the rate of chemical activity in adults. Excess causes thinness and over-activity, and deficiency causes obesity

Endocrine gland	Hormone	Effects
		and sluggishness.
Pancreas (Islets of Langerhans	Insulin	Stimulates conversion of glucose to glycogen in the liver. Deficiency causes diabetes.
Adrenal glands	Adrenalin	Under conditions of 'fight, flight or fright' causes changes which increase the efficiency of the animal. For example, increased heartbeat and breathing, diversion of blood from gut to muscles, coversion of glycogen in the liver to glucose.
Ovaries	Oestrogen	Stimulates secondary sexual characteristics in the female, for example, breast development. Causes the uterus wall to thicken during menstrual cycle.
	Progesterone	Prepares the uterus for implantation.
Testes	Testosterone	Stimulates secondary sexual characteristics in the male, for example, facial hair.

horticulture The science of plant growing, which is of particular importance in food production.

human population curve A diagram showing changes in the human **population**. A population explosion has taken place in the last century due to improvements in food production and public health resulting in increasing **birth rate** and decreasing **death rate**. *See* figure overleaf.

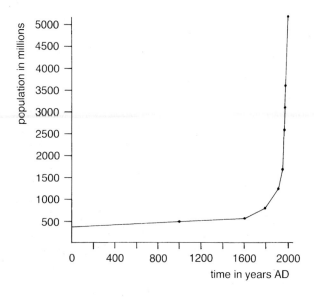

human population curve

humerus A bone of the upper forelimb of **tetrapods**. In humans, this is the bone of the upper arm. *See* **endoskeleton**.

humus The dark-coloured organic material in soil. It consists of decomposing plants and animals, which provides nutrients for plants, and ultimately for animals. *See also* **soil**.

hybrid A plant or animal produced as a result of a cross between two parents of the same **species** that are genetically unlike each other, or between two differing but related species.

hydrolysis The breakdown of complex **organic compounds** by **enzyme** action involving the addition of water. Hydrolysis is the basic reaction of the processes of digestion of proteins, fats, **polysaccharides** and many other compounds:

large
complex compound + H_2O $\xrightarrow{\text{enzyme}}$ small subunits
(starch)   **(glucose)**

hydrotropism **Tropism** relative to water. Plant roots are positively hydrotropic, i.e., they grow towards water.

hypertonic (used of a **solution**) Having a **solute** concentration higher than the medium on the other side of a **selectively permeable membrane**. Such a solution will gain water by **osmosis**. *See also* **hypotonic, isotonic**.

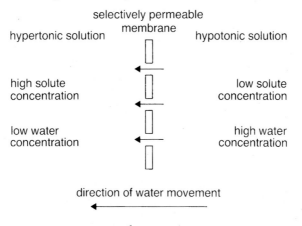

selectively permeable membrane

hypertonic solution

hypotonic solution

high solute concentration

low solute concentration

low water concentration

high water concentration

direction of water movement

hypertonic

hypothesis A suggested solution to a scientific problem. It must be tested by experimentation and, if not validated, must then be discarded. *See also* **scientific method**.

hypotonic (used of a **solution**) Having a **solute** concentration lower than the medium on the other side of a **selectively permeable membrane**. Such a solution will lose water by **osmosis**. *See also* **hypertonic, isotonic**.

I – J

ileum The final region of the mammalian **small intestine**, which receives food from the **duodenum**. The lining of the ileum secretes enzymes that complete the digestion of protein, carbohydrate and fat, into **amino acids**, simple sugars (mainly **glucose**), fatty acids, and glycerol.

Absorption of food occurs in the ileum, which has a large absorbing surface consisting of thousands of finger-like structures called **villi**. The lining of each villus is very thin, allowing the passage of soluble foods. Each villus contains a network of blood capillaries.

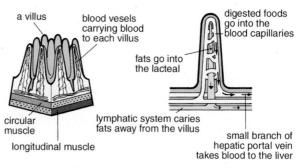

a villus

blood vesels carrying blood to each villus

digested foods go into the blood capillaries

fats go into the lacteal

circular muscle

longitudinal muscle

lymphatic system caries fats away from the villus

small branch of hepatic portal vein takes blood to the liver

ileum *Section through the ileum wall (left) and a section of a villus (right).*

Amino acids and glucose particles diffuse into the blood capillaries from where they are transported first to the liver and then to the general circulation. Fatty acid and glycerol particles pass into the **lacteals** and are circulated via the **lymphatic system**.

Material not absorbed, for example, **roughage**, is passed into the **large intestine**. *See also* **assimilation, digestion**.

imago An adult, sexually mature insect. *See* **metamorphosis**.

immobilization The restriction of movement of cells or enzymes due to their attachment to another substance. For example, enzymes can be fixed onto glass beads in such a way that, having catalysed a required chemical reaction, they can be easily separated from the product and re-used. Immobilization techniques are important in **biotechnology**.

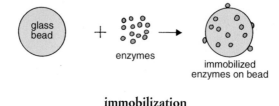

glass bead

+

enzymes

→

immobilized enzymes on bead

immobilization

immunity The ability of the body to resist infectious diseases. Once the body has suffered from a disease and produced specific **antibodies**, these can remain in the blood for months or even years after they helped the organism to recover from the disease. While they are there, they stop the organism catching the same disease again. Those who have antibodies against a disease, or can make them, are *immune* from that disease.

There are different types of immunity. *Active immunity* results when an organism makes its own antibodies as a result of contact with an **antigen**. The antigen may be a **pathogen** or its product. Active immunity usually develops slowly, yet it may last for many years.

Active immunity may be induced artificially by the introduction of antigens into the body. Such antigens may be composed of living, dead or weakened pathogens, which are used to stimulate antibody production but not cause the disease. Suspensions of these pathogens are called **vaccines**.

Passive immunity is given when antibodies produced by active immunity in one organism are transferred to another. The antibodies are extracted from the **blood serum** of a person or animal (usually a horse), which has recovered from the disease. Since no antigen is introduced, there is no stimulation of antibody production; hence passive immunity does not last long, but it does provide immediate protection, unlike active immunity, which requires time for development. The introduction of existing antibodies is done if the body is infected with a disease that is too dangerous to leave until the body's natural defences are activated, or as a preventative measure to guard against such an infection. *See also* **immunization**.

immunization The prevention of infection by artificially introducing a **pathogen** into the body. This stimulates **antibody** production and causes the organism to become immune to the pathogen, which in some cases lasts a lifetime. *See* **immunity**.

In Britain, children are immunized against polio, diphtheria, tetanus, whooping cough and tuberculosis. Girls are also immunized against German measles, a disease that can harm an unborn child. People going abroad may be immunized against diseases such as yellow fever.

implantation Attachment of a mammalian **embryo** to the **uterus** lining at the start of **pregnancy**. In preparation for implantation, the uterus wall thickens with new cells and an increased blood supply. *See also* **fertilization**.

incisors Chisel-shaped cutting **teeth** at the front of the mouth used for biting off pieces of food. *See* **carnivore**, **dental formula**, **dentition**, **herbivore**, **omnivore**.

incomplete dominance A genetic condition in which neither of a pair of **alleles** is dominant but instead they 'blend' to produce an intermediate trait. *See also* **monohybrid inheritance**.

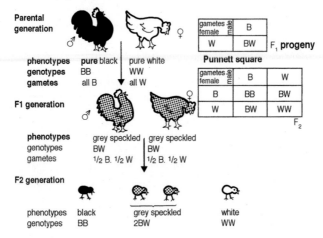

gametes female		B
	male	
	W	BW

F_1 **progeny**

Punnett square

gametes female		B	W
	male		
	B	BB	BW
	W	BW	WW

F_2

Parental generation	phenotypes	**pure** black	pure white
	genotypes	BB	WW
	gametes	all B	all W

F1 generation	phenotypes	grey speckled	grey speckled
	genotypes	BW	BW
	gametes	1/2 B. 1/2 W	1/2 B. 1/2 W

F2 generation	phenotypes	black	grey speckled	white
	genotypes	BB	2BW	WW

incomplete dominance *The inheritance of feather colour in Andalusian fowls.*

incubator 1. A heated container used to grow microorganisms on nutrient agar plates (*see* **agar**). The usual incubation period is 48 hours at 37 °C.
   2. A heated chamber used to maintain the body temperature of premature babies.

indicator organism An organism that can survive only in certain environmental conditions, and hence one whose presence provides information about the environment in which it is found. For example, the bacterium *Escherichia coli* lives in animal gut and is always present in **faeces**. Although the levels of *E. coli* may be harmless, its presence in water indicates sewage **pollution**.

inherited diseases Disorders that are hereditary, i.e. they are passed from generation to generation by **genes**. Examples are **haemophilia**, sickle-cell anaemia, cystic fibrosis and Huntington's chorea.
   Inherited disorders are due to rare **genotypes** producing disease-susceptible **phenotypes**.

inorganic compounds Chemical substances that are derived from the physical environment, and which are not organic.
   The most abundant inorganic compound in living cells is water, which is present in amounts ranging from 5 to 90%. The other inorganic

components of cells are **mineral salts** present in amounts ranging from 1 to 5%. *See* **organic compounds**.

insulin   A vertebrate **hormone** secreted by the *islets of Langerhans* in the **pancreas**. Insulin regulates the conversion of **glucose** to **glycogen** in the **liver**. If the concentration of blood glucose is high, the rate of secretion of insulin is high, and thus glucose is rapidly converted to liver glycogen. If the concentration of blood glucose is low, less insulin is secreted. This is an example of *feedback regulation* found in relation to many hormones. People who suffer from diabetes produce insufficient insulin to control the delicate glucose level balance in their bodies.

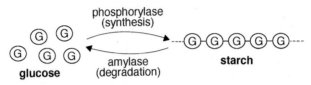

insulin *Feedback regulation of insulin secretion.*

integrated control   The control of **pests** using a combination of chemical and **biological control**.

integument   **1.** The external protective covering of an animal, e.g. skin, cuticle.
   2. The protective layer around flowering plant ovules that, after fertilization, forms the **testa**.

intercostal muscles   The muscles positioned between the ribs of mammals. They are important in lung ventilation. *See also* **breathing**.

intestine   The region of the **alimentary canal** between the **stomach** and the **anus** or **cloaca**. In vertebrates it is the major area of digestion and absorption of food, and is usually differentiated into an anterior **small intestine** and a posterior **large intestine**. *See also* **digestion**.

in vitro   (used of a process or experiment) Conducted outside an organism, for example in a test tube.

in vivo   (used of a process or experiment) Conducted within living organisms.

involuntary muscle or smooth muscle   The type of **muscle** associated with internal tissues and organs in mammals, for example, the gut and blood vessels. It is called 'involuntary muscle' because it is not directly controlled by the will of the organism. Involuntary muscle actions include contraction and dilation of the pupil in the eye, and **peristalsis**. *See also* **antagonistic muscles, voluntary muscle**.

ion An electrically charged **atom** or group of atoms.

iris The structure in the vertebrate **eye** that controls the size of the pupil and hence the amount of light entering the eye.

irritability *See* **sensitivity**.

isotonic (used of a **solution**) Having a **solute** concentration equal to that of the medium on the other side of a **selectively permeable membrane**. There is no water movement across the membrane by **osmosis**. *See also* **hypertonic, hypotonic**.

joint The point in a **skeleton** where two or more **bones** meet and movement may be possible. Moveable joints in mammals are of three types:

(a) *ball and socket joints*, which allow movement in several planes;
(b) *hinge joints*, which allow movement in only one plane;
(c) *gliding joints*, which occur when two flat surfaces glide over one another, allowing a small amount of movement only.

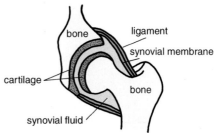

joint *Ball and socket joint.*

   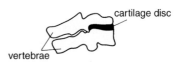

joint *Hinge joint.*　　　　**joint** *Gliding joint.*

K

keratin Strong, fibrous protein present in vertebrate **epidermis** that forms the outer protective layer of skin and also hair, nails, wool, feathers and horns.

key (in biology) A device for identifying unfamiliar organisms. The two types of key shown can be used to identify the major divisions of living organisms:

(a) *branching key*: start at the top and follow the branches by deciding which descriptions best fit the organism to be identified (*see overleaf*);

(b) *numbered key*: at each stage, there is a pair of alternative statements, only one of which can apply to the specimen to be identified. The correct alternative leads to the next choice and so on until the organism is named.

This type of key is preferable to the branching type which can take up too much space.

1a	Multicellular, usually with chlorophyll	Go to 2
b	Multicellular, without chlorophyll, able to move	Go to 3
2a	Plants that reproduce using flowers	Go to 4
b	Plants that reproduce without flowers	Non-flowering plants
3a	Animals with backbones	Go to 5
b	Animals without backbones	Invertebrates
4a	Narrow-leaved plants with one cotyledon in their seeds	Monocotyledons
b	Broad-leaved plants with two cotyledons in their seeds	Dicotyledons
5a	Hair present	Mammals
b	Hair absent	Go to 6
6a	Feathers present	Birds
b	Feathers absent	Go to 7
7a	Dry, scaly skin	Reptiles
b	Other type of skin	Go to 8
8a	Four limbs, moist scaleless skin	Amphibians
b	Limbless	Fish

Using the above key, it is now possible to do a simple classification of organisms such as an ape, pike, newt, oak, snake, grass and earthworm.

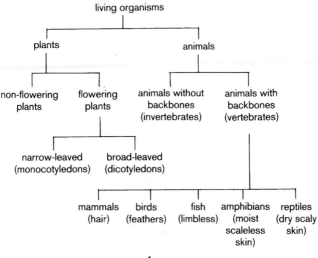

key

kidney The organ of **excretion** and **osmoregulation** in vertebrates, consisting of units called **nephrons**. In humans, the kidneys are a pair of red-brown oval structures at the back of the **abdomen**.

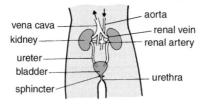

kidney *The position of the kidneys and associated organs in the human body.*

Oxygenated blood enters each kidney via the renal artery, and the renal vein removes deoxygenated blood. Another tube, the **ureter**, connects each kidney with the **bladder**.

The renal artery divides into numerous *arterioles*, which terminate in tiny knots of blood capillaries called **glomeruli**. Each glomerulus (there are

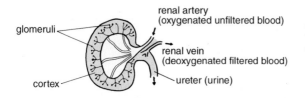

kidney *Section through a kidney.*

about one million in a human kidney) is enclosed in a cup-shaped organ called a **Bowman's capsule**.

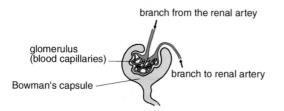

kidney *Glomerulus in Bowman's capsule.*

Two processes occur in the kidneys:

(a) *ultrafiltration*: the vessel leaving each glomerulus is narrower than the vessel entering, causing the blood in the glomerulus to be under high pressure, which forces the blood components with smaller molecules through the selectively permeable capillary wall into the Bowman's capsule (*see* **selectively permeable membrane**).

Large particles (unfiltered)	Small particles (filtered)
Blood cells	Glucose
Plasma proteins	Urea
	Mineral salts
	Water
	Amino acids

(b) *reabsorption*: the fluid filtered from the blood (*filtrate*) passes from the Bowman's capsule down the renal tubule where reabsorption of useful materials occurs, i.e all the glucose and amino acids, and some of the salts and water are reabsorbed into the blood.

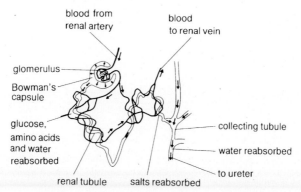

kidney *Reabsorption of useful substances.*

The liquid remaining after filtration and reabsorption by the kidney is a solution of salts and urea in water (*urine*) and is passed to the **bladder** via the ureters from where it is expelled via the **urethra** under the control of a **sphincter muscle**. *See also* **antidiuretic hormone**.

kidney machine An artificial **kidney** that purifies the blood of a person with diseased kidneys. Blood is taken from the person's arm artery and passed through a **selectively permeable membrane** where **urea** and excess **mineral salts** are removed into a rinsing solution. This process is called *dialysis*. The purified blood is then returned to the person's arm vein.

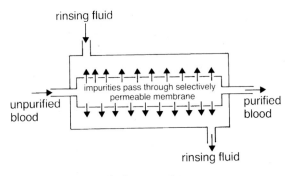

rinsing fluid

impurities pass through selectively permeable membrane

unpurified blood

purified blood

rinsing fluid

kidney machine

kilojoule The unit used to measure the energy content of food. *See also* **food calorimeter**.

kingdom Any of the five great divisions of living organisms, i.e. the Animal, Plant, **Protista**, **Fungi** and **Bacteria** kingdoms. *See also* **classification**.

L

lacteals **Lymph** vessels within the **villi** of the vertebrate **intestines**. The products of the digestion of fats (fatty acids and glycerol) diffuse into the lacteals and are circulated via the **lymphatic system**.

lactic acid An organic acid ($CH_3CHOHCOOH$) produced during **respiration** in many animal cells, including vertebrate muscle cells and certain bacteria. *See also* **oxygen debt**.

large intestine The posterior region of the vertebrate **intestine**. In humans, at the entry to the large intestine there is a region called the **caecum**, from which projects the **appendix**, but most of the large intestine consists of the **colon**, which leads to the **rectum**. The large intestine receives undigested material from the **ileum**. *See also* **digestion**.

larva An intermediate, sexually immature stage in the **life history** of some animals between hatching from the egg and becoming adult. Examples are amphibian tadpoles and butterfly caterpillars. *See* **metamorphosis**.

larynx A region at the upper end of the **trachea** of tetrapods opening into the **pharynx** and specialized to close the **glottis** during swallowing.
  In mammals, amphibians and reptiles, vocal cords within the larynx produce sound. *See* **breathing**.

leaching The removal of **mineral salts** and other dissolved substances from the soil by the downward movement of water, mainly rain. The substances are carried into streams and rivers.

leaf That part of a flowering plant which grows from the **stem** and is typically flat and green. The functions of leaves are:
(a) **photosynthesis**;
(b) **gas exchange**;
(c) **transpiration**.
  The cells immediately beneath the upper **epidermis** are called **palisade mesophyll** cells.

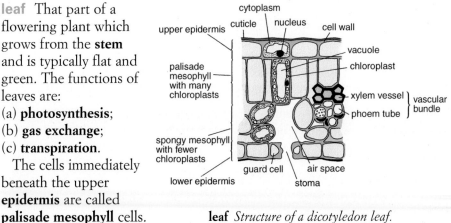

leaf *Structure of a dicotyledon leaf.*

They are closely packed and contain many **chloroplasts**, most of which are situated at the upper part of the cell, thus obtaining maximum light. Between the palisade cells and the lower epidermis are the **spongy mesophyll** cells, which contain fewer chloroplasts. These cells are loosely packed, being separated by air spaces, which allow free circulation of air between the leaves and the atmosphere via the **stomata**.

The positioning of most of the stomata on the lower epidermis means that the important palisade layer is not interrupted by too many air spaces.

Water for photosynthesis is transported in **xylem** vessels in the mid-rib and veins, while the synthesized **carbohydrate** is transported throughout the plant via the **phloem** sieve tubes.

lens A transparent structure behind the pupil of the vertebrate **eye**, important in focusing the image on the **retina**, and in **accommodation**.

lenticel One of many pores developing in woody **stems** and **roots** when **epidermis** is replaced by bark, through which **gas exchange** occurs.

leucocyte *See* **white blood cell**.

lichen A plant formed by **mutualism** between an **alga** and a **fungus**. The alga supplies carbohydrate and oxygen to the fungus, and receives water and **mineral salts** in return.

life history or life cycle The various stages of development that organisms undergo from egg to adult. *See also* **metamorphosis**.

ligament A strong band of **collagen** connecting the **bones** at moveable vertebrate **joints**. Ligaments strengthen the joint, allowing movement in only certain directions and preventing dislocation.

lignin An organic compound deposited in the cell walls of **xylem** vessels, giving strength. Lignin is an important constituent of wood.

limiting factor Any factor of the **environment** whose level at a particular time inhibits some activity of an organism or **population** of organisms.

lipase An **enzyme** that digests **fat** into fatty acids and glycerol by **hydrolysis**. In mammals, lipase is secreted by the **pancreas** and the **ileum**. *See* diagram opposite.

lipid *See* **fat**.

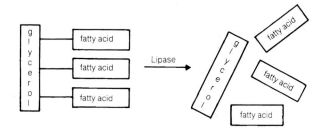

lipase *Hydrolytic breakdown of fat.*

liver The largest organ of the vertebrate body, occupying much of the upper part of the **abdomen**, in close association with the **alimentary canal**. *See also* **digestion, circulatory systems**.

Some of the many functions of the liver are:
(a) production of **bile**;
(b) **deamination** of surplus **amino acids**;
(c) regulation of blood sugar by interconversion of **glucose** and **glycogen**;
(d) storage of iron, and **vitamins** A and D;
(e) detoxification of poisonous by-products;
(f) release and distribution of heat produced by the chemical activity of liver cells;
(g) conversion of stored fat for use by the tissues;
(h) manufacture of **fibrinogen**.

long sight or **hypermetropia** A human eye defect mainly caused by a shorter than normal distance from the **lens** to **retina**. This results in near objects being focused behind the retina giving blurred vision. Long sight is corrected by wearing converging (*convex*) lenses.

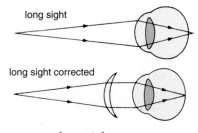

long sight

lungs The organs of **breathing** in mammals, amphibians, reptiles and birds. In mammals, the lungs are two elastic sacs in the **thorax** that can be expanded or compressed by movements of the thorax in such a way that air is continually taken in and expelled. The trachea (windpipe) connects the lungs with the atmosphere. It divides into two **bronchi**, which enter the lungs and further divide into many smaller *bronchioles*. The bronchioles terminate in millions of air-sacs called **alveoli**, which are the **gas exchange** surfaces and which are in close contact with blood vessels, bringing blood from and to the heart.

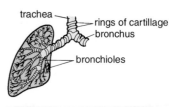

lungs *Air passages in the lung.*

luteinizing hormone (LH) A **hormone** secreted by the vertebrate **pituitary gland**. *See* **ovulation**.

lymph Fluid drained from blood **capillaries** in vertebrates as a result of high pressure at the arterial end of the capillary network. Lymph or *tissue fluid*, which is similar to **plasma** (except for a much lower concentration of **plasma proteins**) bathes the tissues and acts as a medium in which substances are exchanged between capillaries and cells. For example, oxygen and **glucose** diffuse into the cells while carbon dioxide and **urea** are removed. Lymph drains back into capillaries or into vessels called lymphatics, which then connect with the general circulation via the **lymphatic system**.

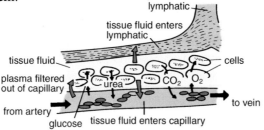

lymph *The lymphatics, capillaries and cells.*

lymphatic system A system of fluid-containing vessels (*lymphatics*) in vertebrates, which return lymph to the general blood circulation. The lymphatic system is also important in:
(a) transporting the products of fat digestion;
(b) production of **white blood cells** and **antibodies**.

lymph nodes Structures within the **lymphatic system** that filter bacteria from **lymph** and produce **white blood cells** and **antibodies**.

lymphocyte A type of **white blood cell** of vertebrates. There are two types of lymphocytes:

B-lymphocytes, which are derived from the bone marrow. These cells make and release **antibodies** into the bloodstream to attack **antigens**.

T-lymphocytes, which are derived from the thymus gland. These make antibodies, which remain attached to the surface of the cell. It is therefore the lymphocyte itself that attacks the antigens. There are several sorts of T-lymphocyte. Some of them destroy cancer cells; others recognize and attack infected body cells and foreign tissue that has been grafted into the body. These are called the *killer T-cells*.

Some B- and T-lymphocytes, having made a specific antibody, become *memory cells*. These remain in the body ready to make the same antibody if it is needed again. This provides a very rapid response to any subsequent invasion by similar antigens.

M

malnutrition Inadequate nutrition caused by the lack of a **balanced diet** or by insufficient food. In tropical countries, diseases associated with protein and **carbohydrate** deficiency cause many deaths among young children.

medulla The central part of a tissue or organ such as the mammalian kidney. *See also* **cortex**.

medulla oblongata The posterior region of the vertebrate **brain**. It is continuous with the **spinal cord** and in mammals controls heartbeat, breathing, **peristalsis** and other involuntary actions.

meiosis A method of *nuclear division* that occurs during the formation of **gametes** when a **diploid nucleus** gives rise to four **haploid** nuclei. There are two consecutive divisions in the process.

metaphase:
nuclear spindle
forms and
chromosomes
move to the equator

anaphase:
centromeres repel
each other carrying
chromosomes towards
the poles of the spindle

telophase:
nuclear membranes
form around
the groups
of chromosomes

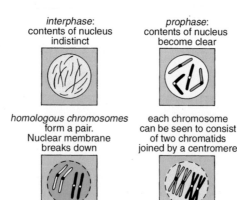

meiosis *Stages in the two divisions of the process.*

interphase:
contents of nucleus
indistinct

prophase:
contents of nucleus
become clear

homologous chromosomes
form a pair.
Nuclear membrane
breaks down

each chromosome
can be seen to consist
of two chromatids
joined by a centromere

meiosis *Stages in the two divisions of the process.*

During the first division of meiosis the **homologous chromosomes** may exchange genetic material as they lie side by side. This leads to **variation** in the resulting nuclei. This process is called *crossing over.*

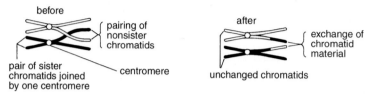

before

pairing of
nonsister
chromatids

pair of sister
chromatids joined
by one centromere

centromere

after

exchange of
chromatid
material

unchanged chromatids

meiosis *Crossing over between non-sister chromatids.*

menopause The cessation of the **menstrual cycle**. It signals the end of the reproductive period in human females.

menstrual cycle The reproductive cycle of female primates (monkeys, apes, humans). This cycle is under the control of **hormones**. In human females, the cycle lasts about 28 days, during which the **uterus** is prepared for **implantation**. If **fertilization** does not occur, the new uterus lining and unfertilized **ovum** are expelled, which results in bleeding from the **vagina** (**menstruation**). *See also* **ovulation**.

menstruation The discharge of **uterus** lining-tissue and blood from the **vagina** in human females at the end of a **menstrual cycle** in which **fertilization** has not occurred.

messenger RNA *See* **RNA**, **protein synthesis**.

metabolic water One of the products of aerobic **respiration**. It is an important source of water for desert animals.

metabolism The sum of all the physical and chemical processes occurring within a living organism. These include both the synthesis (*anabolism*) and breakdown (*catabolism*) of compounds. *See also* **basal metabolic rate**.

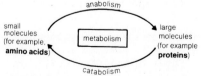

anabolism

small
molecules
(for example,
amino acids)

metabolism

large
molecules
(for example
proteins)

catabolism

metabolism *The synthesis and breakdown of compounds within an organism.*

metamorphosis The period in the **life history** of some animals when the juvenile stage is transformed into an adult.

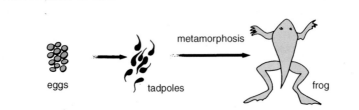

metamorphosis *Amphibians go through a tadpole stage.*

Incomplete metamorphosis is a type of development in which there are relatively few changes from juvenile form to adult. It occurs in insects such as the dragonfly, locust and cockroach, in which the juvenile form (**nymph**) resembles the adult except that it is smaller, wingless, and sexually immature.

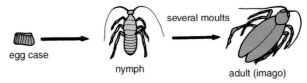

metamorphosis *Incomplete metamorphosis in the cockroach.*

Complete metamorphosis involves great changes from **larva** to adult. It occurs in insects such as butterfly, moth, housefly, etc., and the larvae in such life histories are maggots, grubs or caterpillars (depending on the species) and are quite unlike the adult form. A series of moults (**ecdyses**) produces the **pupa**, which then becomes completely reorganized and develops into the adult, the only sexually mature stage.

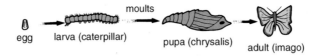

metamorphosis *Complete metamorphosis in a butterfly.*

microbe   An alternative term for **microorganism**.

microbiology   The study of **microorganisms**.

microbiology experiments   The use of sterile nutrient agar plates (*see* **agar**) to culture microorganisms. The diagrams (*see opposite*) show that microorganisms are found in various habitats including air, soil, water, exposed surfaces, food and on other living organisms.

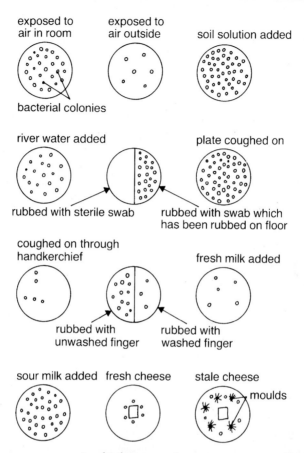

exposed to
air in room

exposed to
air outside

soil solution added

bacterial colonies

river water added

plate coughed on

rubbed with sterile swab

rubbed with swab which
has been rubbed on floor

coughed on through
handkerchief

fresh milk added

rubbed with
unwashed finger

rubbed with
washed finger

sour milk added   fresh cheese

stale cheese

moulds

microbiology experiments

microorganisms Very small living organisms that can usually only be seen with the aid of a **microscope**. Microorganisms include **protozoa, algae, viruses, fungi** and **bacteria**.

microorganisms and food Most foods contain **microorganisms**, which are usually destroyed during cooking. Some foods must be treated before consumption, an example being milk, which is heated at 80 °C for 30 seconds. This is called *pasteurization*.

Microorganisms cause food to go off, and various methods are used to preserve food. These include:

(a) *freezing*: low temperatures inhibit microorganism growth;

(b) *salting*: this causes water to be drawn from microorganisms resulting in so-called osmotic death;

(c) *dehydration*: this deprives microorganisms of necessary water;

(d) *smoking*: this adds chemicals that kill microorganisms;
(e) *high temperature sterilization*: this is used in the canning of foods;
(f) *chemical preservatives*: e.g. an **acid** such as acetic acid (vinegar), which is added to some foods to cause an acid **pH** unfavourable to microorganisms.

micropyle 1. A pore in a **seed** through which water is absorbed at the start of **germination.**
   2. A pore in the **ovule** of a flower through which the **pollen** tube delivers the male **gamete.**
   3. A pore in the **ovum** of insects through which the **spermatozoon** enters.

microscope An instrument used to magnify structures, for example, cells or organisms, which are not visible to the naked eye:
(a) *light microscope*: light illuminates the specimen which is magnified by glass lenses. The magnification of the microscope is found by multiplying the magnification of the objective lens (e.g. × 40) by the magnification of the eyepiece lens (e.g. × 10) to give the total magnification (in this example × 400). The maximum possible magnification using a light microscope is × 1500. Thin specimens are placed on a glass slide and may be stained with dyes that show up particular structures;

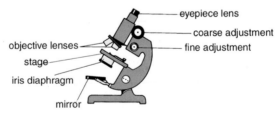

microscope *Parts of a light microscope.*

(b) *phase-contrast microscope*: this allows the viewing of transparent and unstained structures;
(c) *electron microscope*: the most advanced type, giving magnification as high as × 500,000.

milk teeth or deciduous teeth The first set of **teeth** occurring in most mammals. For example, humans have 20 milk teeth, which are replaced during childhood by the larger permanent teeth.

mineral salts Components of soil formed from rock **weathering** and **humus** mineralization, and found in solution in soil water. Mineral salts are absorbed by plant roots and transported through the plant in the **transpiration stream**. Like **vitamins**, mineral salts are required in tiny amounts, but are nevertheless vital for plant and ultimately animal nutrition; the absence of a particular mineral salt can lead to mineral deficiency disease and death. Plants require at least twelve mineral salts for healthy growth:

(a) *essential elements* required in relatively large quantities: nitrogen, phosphorus, sulphur, potassium, calcium, magnesium;

(b) *trace elements* required in very small amounts: manganese, copper, zinc, iron, boron, molybdenum.

Some mineral salts are required by plants, some are required by animals and some by both.

Mineral salt	Function	Some effects of deficiency
Phosphorus	Components of ATP, nucleic acids, cell membrane, animal bones	Stunted plant growth
Calcium	Component of plant cell walls and animal bones	Rickets in humans
Nitrogen	Component of protein and nucleic acids	Poor reproductive development in plants
Iron	Component of haemoglobin	Anaemia in humans
Magnesium	Component of chlorophyll	Pale yellow plant leaves (chlorosis)

mitochondrion A microscopic cell **organelle** in the **cytoplasm** of aerobic cells, in which some **respiration** reactions occur. The inner membrane of a mitochondrion wall is highly folded, giving rise to a series of partitions called *cristae*, which greatly increase the surface area for the attachment of respiratory enzymes. *See also* **aerobe, enzymes.**

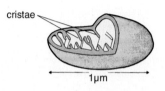

cristae

1μm

mitochondrion
Section showing cristae.

mitosis The process necessary for the growth of an organism. The
nucleus of a **cell** divides in such a way that the resultant *daughter cells*
receive precisely the same numbers and types of **chromosomes** as the
original *mother cell*. During this type of nuclear division the chromosomes
of the mother cell (the dividing cell) are first duplicated and then passed in
identical sets to the two daughter cells.

In stage 1, Chromosomes appear as coiled threads. Two structures called
centrioles are seen outside the nucleus.

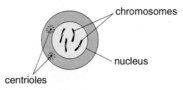

mitosis *Stage 1.*

Each chromosome makes a duplicate of itself. These duplicates or
chromatids are joined at a *centromere*.

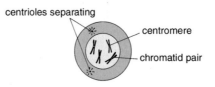

mitosis *Stage 2.*

The *nuclear membrane* disappears. The centrioles move to opposite poles
of the cell and produce a system of fibres called a *spindle*. The chromatid
pairs line up at the equator of the cell.

mitosis *Stage 3.*

The chromatid pairs separate and move to opposite poles of the cell.

mitosis *Stage 4.*

New nuclear membranes form and the **cytoplasm** starts to divide.

mitosis *Stage 5.*

The chromosomes disappear and the cells return to the resting state.

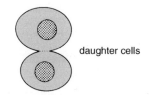

daughter cells

mitosis *Stage 6.*

molars Broad, crowned, grinding **teeth** at the sides and back of the mouth, used for crushing food prior to swallowing. Together with **premolars**, these teeth are also known as cheek teeth. They are found in **omnivores** and **herbivores** but are replaced by **carnassial** teeth in **carnivores**. *See also* **dental formula, dentition**.

molecule The smallest complete part of a chemical compound that can take part in a reaction. Within a molecule, atoms occur in fixed proportions.

monocotyledons The smaller of the two subsets of flowering plants, the other being **dicotyledons**. The characteristics of monocotyledons are:
(a) one **cotyledon** in the seed;
(b) parallel veins in leaves;
(c) narrow leaves;
(d) scattered **vascular bundles** in stem;
(e) flower parts usually in threes or in multiples of threes.
Examples are cereals and grasses.

monohybrid inheritance The inheritance of one pair of contrasting characteristics.
  For example, in the fruit-fly *Drosophila*, one variety is **pure-breeding** normal-winged, and another is pure-breeding vestigial-winged, the normal-winged **allele** being **dominant** to the vestigial-winged allele.

The approximate ratio in the F_2 generation is:

$$\frac{\text{normal}}{\text{vestigial wing}}=\frac{3}{1}$$

When normal-winged is crossed with vestigial-winged, the vestigial trait seems to disappear, only to reappear again to a limited extent in the next generation, suggesting that the **F_1 generation** must have possessed this trait without showing it. A trait such as normal wing which always appears in a cross between contrasting parents is described as **dominant**, while a trait such as vestigial wing, which is 'lost' in F_1 generation **progeny**, apparently masked by a dominant trait, is called **recessive**.

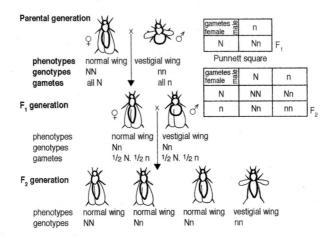

monohybrid inheritance *The results of crossing normal-winged female Drosophila with vestigial-winged males.*

For each trait, an organism receives one **gene** from the male **gamete** and one from the female gamete. The resulting organism, i.e. the **zygote**, contains two genes for every trait, but the gametes contain only one (as a result of **meiosis**). If the paired genes for a particular trait are identical, the organism is said to be **homozygous** or *pure* for that trait. When an organism has two different genes for a trait, it is described as **heterozygous** or **hybrid**. Alternative forms of a gene are described as *allelomorphs* or alleles. Hence, if the allele for normal wing is represented by N and that for vestigial wing by n:

NN is homozygous normal wing
Nn is heterozygous normal wing
nn is homozygous vestigial wing

Organisms that are homozygous for a particular trait produce only one type of gamete for that trait, while heterozygous organisms produce two gamete types. The results of **fertilization** can then be worked out using a **Punnett square**. *See* **backcross, incomplete dominance**.

monosaccharides Single **sugar** carbohydrates, which are the subunits of more complex carbohydrates and named on the basis of the number of carbon atoms present. For example:

$$C_3H_6O_3 \qquad C_5H_{10}O_5 \qquad C_6H_{12}O_6$$
triose sugar   pentose sugar   hexose sugar

Hexose sugars are common carbohydrates and include **glucose**.

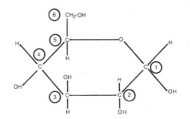

monosaccharides *Structure of glucose (C6H12O6).*

motile (used of organisms or parts of organisms) Able to move.

mucus A slimy fluid secreted by **goblet cells** in vertebrate **epithelia**. Mucus traps dust and bacteria in mammalian air passages, lubricates the surfaces of internal organs, and facilitates the movement of food through the gut while preventing the digestive enzymes from reaching and digesting the gut itself.

multicellular (used of an organism) Consisting of many **cells**. Most animals and plants are multicellular. *Compare* **unicellular**.

muscle Animal tissue consisting of cells that are capable of contraction as a result of **nerve impulses**, thus producing movement, both of the organism as a whole and of internal organs. *See also* **antagonistic muscles, involuntary muscles, voluntary muscles**.

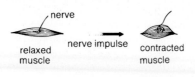

muscle *Contraction.*

mutagenic agents Factors that speed up the rate of **mutation**. These include certain chemicals such as mustard gas; irradiation with ultraviolet rays or X-rays; and atomic radiation.

mutation A change in the structure of **DNA** in **chromosomes**. Mutations occur rarely; when they occur in the **gametes**, or the cells that give rise to them, they are inheritable and most confer disadvantages on the organisms inheriting them. Mutations can result in beneficial **variations** within a population, which can lead to **evolution**, and although they occur spontaneously, they can also be induced by exposure to excessive radiation or other **mutagenic agents**.

mutualism A **symbiotic** relationship in which both organisms benefit. For example, the **intestinal bacteria** of **herbivores** digest the **cellulose** of plant cell walls, the products of which are then used by the herbivore.

mycorrhiza A **symbiotic** association between **fungi** and the **roots** of certain plants. The fungi provides the plants with **amino acids** and in return receives **carbohydrates**.

N

nastic movement A growth response by plants to a stimulus that is independent of the direction of the stimulus. An example is the opening and closing of a flower in response to light intensity. *See also* **tropism**.

natural selection The theory proposed by Charles Darwin to explain how **evolution** could have taken place. Darwin suggested that individuals in a **species** differ in the extent to which they are adapted to their environment. Thus, in **competition** for food, etc., the better-adapted organisms will survive, and pass on their favourable **variations**, while the less well adapted will be eliminated.

Over a large number of generations, natural selection can change the characteristics of a species and contribute to the development of new species. All the species that exist today are thought to have evolved, by **mutation** and natural selection, from simple organisms which first developed millions of years ago.

nephron A subunit of the vertebrate **kidney**, consisting of a **Bowman's capsule**, **glomerulus**, and renal tubule. *See also* **excretion**, **osmoregulation**.

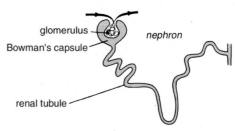

nephron *Section through a nephron.*

nerve impulses The electrical messages by which information is transmitted rapidly throughout **nervous systems**. Nerve impulses are initiated at **receptor** cells as a result of **stimuli** from the environment. In vertebrates, the impulses are conducted to the **central nervous system**, where they trigger other impulses that are relayed to **effector** organs. *See also* **neurones**, **synapse**.

nervous system A network of specialized cells in **multicellular** animals, which acts as a link between **receptors** and **effectors**, and thus coordinates the animal's

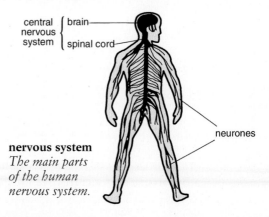

nervous system *The main parts of the human nervous system.*

activities. In mammals, the nervous system consists of the **brain** and **spinal cord** (which together form the **central nervous system**) and **neurones** connecting all parts of the body.

neural (used of functions and parts of the body) Related to the **nervous system**.

neurones or nerve cells **Cells** that are the basic units of mammalian **nervous systems**.

There are two types of neurone:

(a) *sensory neurones*: these conduct **nerve impulses** from **receptors** to the **central nervous system** (CNS), i.e. from eyes, ears, skin, etc.;

(b) *motor neurones*: these conduct **nerve impulses** from the CNS to **effectors**, such as muscles and **endocrine glands**.

Each neurone consists of three parts:

(a) a cell body containing **cytoplasm** and **nucleus** and forming the *grey matter* in the brain and spinal cord;

(b) fibres that carry nerve impulses into cell bodies. In sensory neurones, this fibre is a single *dendron* while in motor neurones there are numerous *dendrites*. In the CNS such fibres form *white matter*;

(c) fibres called **axons**, which carry nerve impulses from cell bodies.

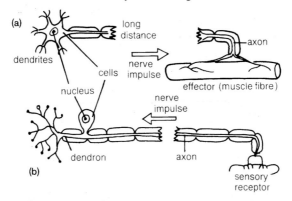

neurones *The structure of (a) motor neurone and (b) sensory neurone.*

neutralize *See* **alkali**.

neutron One of the particles found in the **nucleus** of all **atoms** except hydrogen. It has no charge.

niche The status or way of life of an organism within a **community**. For example, a **herbivore** and a **carnivore** may share the same **habitat** but their different feeding methods mean that they occupy different niches.

nitrification The conversion of organic nitrogen compounds by nitrifying bacteria in the soil, for example, ammonia into nitrates, which can be absorbed by plants. Ammonia is first converted to nitrites by *Nitrosomonas* bacteria and the nitrites to nitrates by *Nitrobacter* species. *See also* **nitrogen cycle**.

nitrogen cycle The circulation of the element nitrogen and its compounds in nature, caused mainly by the **metabolic** processes of living organisms.

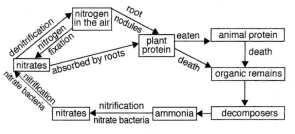

nitrogen cycle *The main steps.*

nitrogen fixation The conversion of atmospheric nitrogen by certain microorganisms into organic nitrogen compounds. Nitrogen-fixing bacteria live either in soil, air, or within the **root nodules** of *leguminous plants* (peas, beans, clover). The activity of these organisms, such as *Azotobacter* and *Rhizobium*, enriches the soil with nitrogen compounds. *See also* **nitrogen cycle, root nodule**.

normal distribution curve A bell-shaped curve obtained when continuous variation (*see* **variation**) is measured in a **population**.

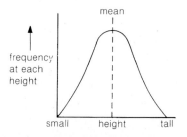

normal distribution curve

nuclear division *See* **meiosis, mitosis**.

nucleic acids **Organic compounds** found in all living organisms, particularly associated with the **nucleus** of the cell and consisting of subunits called *nucleotides*.

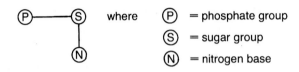

where   (P) = phosphate group

nucleic acids *A polynucleotide chain.*

The sugar group of one nucleotide can combine with the phosphate group of another to form a *polynucleotide* chain. Such polynucleotide chains are the basis of nucleic acid structure. *See also* **DNA, RNA.**

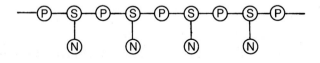

nucleic acids *A polynucleotide chain.*

nucleus A structure within most cells in which the **chromosomes** are located. It is isolated from the **cytoplasm** by a *nuclear membrane.* Chromosomes are visible only during nuclear division. As the chromosomes contain the hereditary information, the nucleus controls all the cell's activities through the action of the genetic material **DNA.**

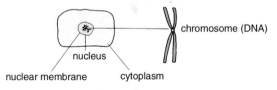

nucleus *The nucleus of a cell and one of its chromosomes.*

nutrient agar plate *See* **agar.**

nymph The juvenile form of certain insects, which resembles the **imago** except that it is smaller, wingless and sexually immature. *See also* **metamorphosis.**

O

oesophagus A region of the **alimentary canal**, connecting the mouth with the digestive areas. In vertebrates, it runs between the **pharynx** and the **stomach** and transports food by **peristalsis**. *See also* **digestion**.

oestrogen A **hormone** secreted by vertebrate **ovaries**, which stimulates the development of **secondary sexual characteristics** in female mammals and is important in the **menstrual cycle**.

olfactory (used of parts of the body and functions) Related to the sense of **smell**.

omnivore An animal that feeds on both plants and animals. Omnivores include humans, whose **dentition**, like other omnivores, contains both **herbivore** and **carnivore** features, consisting of biting, ripping and grinding teeth, which suit the mixed diet. *See also* **dental formula**, **teeth**.

optic (used of parts of the body) Related to the **eye** and its functions.

optic nerve A **cranial nerve** of vertebrates conducting **nerve impulses** from the **retina** of the eye to the brain.

oral (used of parts of the body and functions) Related to the mouth.

oral hygiene The maintenance of healthy conditions in the human mouth in order to reduce tooth decay. Bacteria in the mouth convert food particles stuck to teeth into acid, which attacks the teeth and causes decay. Regular brushing with toothpaste reduces the likelihood of tooth decay.

order A unit used in the **classification** of living organisms, consisting of one or more **families**.

organ A collection of different **tissues** in a plant or animal that form a structural and functional unit. Examples are the liver and a plant leaf.
   Different organs may then be associated together to constitute a *system*, e.g. the digestive system.

organelle A structure found in the **cytoplasm** of **cells**. Examples are **mitochondria**, **chloroplasts**. *See also* **cell differentiation**.

organ of Corti *See* **cochlea**.

organic compounds Compounds containing the element carbon, found in living organisms. The major organic compounds are **carbohydrates**, **fats**,

nucleic acids, **proteins** and **vitamins**. *See also* **inorganic compounds**.

osmoregulation Control of the **osmotic pressure**, and therefore the water content, of an organism. *See* **osmosis**.

In terrestrial organisms, water is gained from food and drink and as a by-product of respiration, and is lost by sweating, in exhaled air and as urine.

Water and mineral salt balance in terrestrial animals is mainly under the control of the kidneys. For osmoregulation in fish, *see diagram*.

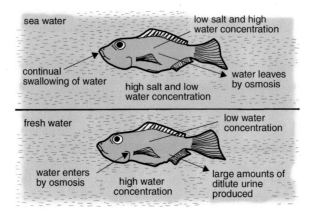

osmoregulation *Water balance in a sea fish (above) and a freshwater fish (below).*

osmosis

The **diffusion** of **solvent** (usually water) particles through a **selectively permeable membrane** from a region of high solvent concentration to a region of lower solvent concentration.

Examples of selectively permeable membranes are the cell membrane (*see* **cell**) and *visking tubing* (used in dialysis – *see* **kidney machine**).

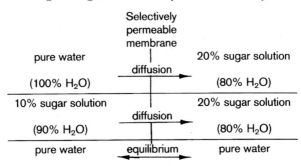

osmosis

Such membranes are thought to have tiny pores that allow the rapid passage of small water particles, but restrict the passage of larger *solute* particles.

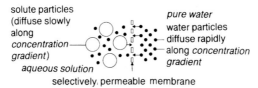

solute particles (diffuse slowly along concentration gradient)

aqueous solution

pure water
water particles diffuse rapidly along *concentration* gradient

selectively permeable membrane

osmosis *Passage of solvent (water) particles through a selectively permeable membrane.*

Since the cell membrane is selectively permeable, osmosis is important in the passage of water into and out of cells and organisms, and depends on **osmotic pressure**. *See also* **turgor, wall pressure**.

osmotic pressure The pressure exerted by the osmotic movement of water, which can be demonstrated in an *osmometer*.

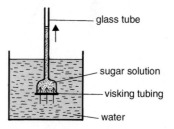

glass tube

sugar solution

visking tubing

water

osmotic pressure
An osmometer.

Water moves by **osmosis** into the sugar solution via the visking tubing, causing the liquid level in the tube to rise. Osmotic pressure depends on the relative **solute** concentrations of the **solutions** involved. The osmotic pressure that a solution is capable of developing is called its *osmotic potential*, but is only realized in an osmometer. *See also* **turgor, wall pressure**.

ossicles The three tiny linked bones in the mammalian middle ear. *See* **ear**.

oval window A membrane separating the middle ear and inner ear in mammals. *See* **ear**.

ovary 1. A hollow region in the **carpel** of a flower, containing one or more **ovules**. *See also* **fertilization**.
2. The reproductive organ of female animals. In vertebrates, there are two ovaries, which produce the **ova** and also release certain sex hormones. *See also* **fertilization, ovulation**.

oviduct A tube in animals that carries **ova** from the **ovaries**. In mammals there are two oviducts leading to the **uterus**, and **fertilization** occurs within the oviduct.

ovulation The release of an **ovum** from a mature **Graafian follicle** on the surface of a vertebrate **ovary**, from where it passes into the **oviduct** and then into the **uterus**.

In human females, ovulation is controlled by hormones from the **pituitary gland**. The sequence of events in the female's reproductive behaviour is called the **menstrual cycle**. **Follicle stimulating hormone** (FSH) induces the maturation of ova and causes the ovaries to produce **oestrogen**. **Luteinizing hormone** (LH) triggers ovulation and also the release of **progesterone** by the ovaries.

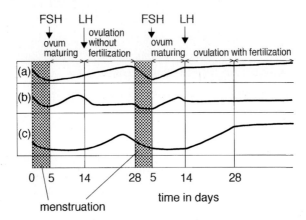

ovulation *The main features of the human menstrual cycle:*
(a) variation in thickness of wall of uterus;
(b) oestrogen level in blood (stimulates repair of the uterus wall);
(c) progesterone level in blood (prepares uterus for implantation).

If the mature ovum is not fertilized, it is expelled with the new uterus lining and some blood via the **vagina**, a process called **menstruation**.

ovule A structure in flowering plants, which develops into a **seed** after fertilization. *See also* **carpel**.

ovum (*pl.* **ova**) An unfertilized female **gamete** produced in the **ovary** of many animals and containing a **haploid** nucleus. *See also* **fertilization**, **meiosis**, **ovulation**.

oxygenated blood *See* **artery**.

oxygen debt A deficit of oxygen that occurs in **aerobes** when work is done with inadequate oxygen supply. For example, in mammalian muscle during exercise, the oxygen supply may be insufficient to meet the energy demand. When this happens, the cells produce energy by **anaerobic respiration**, **lactic acid** being a by-product.

The accumulation of lactic acid causes muscle fatigue but is eventually reduced as oxygen intake returns to normal after the period of exercise. This shortfall of oxygen must be repaid by increased oxygen intake (panting). *See also* **blood**, **respiration**.

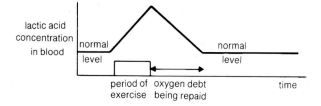

oxygen debt *The effect of exercise on the lactic acid concentration of the blood.*

ozone layer A layer of the atmospheric gas ozone (O_3), which surrounds the Earth. It is thought that the layer is being depleted as the result of chlorofluorocarbons (CFCs), which are gases released by aerosol cans. This may allow more ultraviolet (UV) radiation from the Sun to reach the Earth's surface, causing skin cancer and disturbing weather patterns.

P – Q

pacemaker **1.** A structure in the right **atrium** of the **heart**, which initiates the **heartbeat**.
2. An electronic device implanted in the heart to stimulate and regulate the heartbeat.

palaeontology *See* **fossil**.

palisade mesophyll The main tissue carrying out **photosynthesis** in the leaf, situated below the upper **epidermis** and containing many **chloroplasts**.

pancreas A gland situated near the **duodenum** of vertebrates. It releases an alkaline fluid into the duodenum, containing digestive enzymes, including **lipase**, amylase and **trypsin**. The pancreas also contains tissue known as the *islets of Langerhans*, which secretes the hormone **insulin**.

parasite An organism that feeds in or on another living organism, which is called the *host*, and which does not benefit and may be harmed by the relationship. Parasites of man include fleas, lice and tapeworms. *See also* **ectoparasite, endoparasite**.

parasitism *See* **symbiosis**.

parental generation The first organisms crossed in a breeding experiment, producing progeny known as the F_1 **generation**. *See* **monohybrid inheritance**.

patella A bone over the front of the knee joint in many vertebrates. In humans it is the kneecap.

pathogen A term used to describe an organism causing disease in another species. Examples are **viruses**, **bacteria** and tapeworms. *See also* **parasite**.

pathogenic microorganisms **Microorganisms** that cause disease. Such organisms enter the body by various routes: contaminated food, inhaled air, insect bites and skin wounds. However, the body has mechanisms for preventing infection:
(a) the acid **pH** of the stomach kills many organisms;
(b) the formation of **blood clots** at wounds restricts entry;
(c) **white blood cells** destroy microorganisms by **phagocytosis**;
(d) **antibodies** in the blood neutralize the poisons produced
      by microorganisms;
(e) **cilia** and **mucus** clear the air passages.

If these natural defence mechanisms are overcome, the infection can then be treated using **antibiotics**. Prevention is also possible by **immunization**.

pectoral Relating to that part of the body at the **anterior** end of the trunk of an animal (e.g. the shoulders) to which the forelimbs are attached.

pelvic Relating to that part of the body of an animal that forms the lower abdomen and to which the hindlimbs are attached.

penicillin The first **antibiotic**, discovered by Alexander Fleming in 1928.

penis An organ in mammals by which the male **gametes** (**spermatozoa**) are introduced into the female body. It also contains the **urethra** through which **urine** is discharged. *See also* **fertilization**.

pentadactyl limb A limb with five digits, characteristic of **tetrapod** animals. There is a basic arrangement of bones, which is modified in many species.

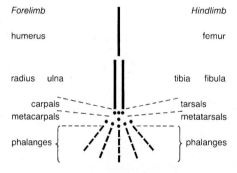

pentadactyl limb

pepsin A **protease enzyme** secreted by the wall of the vertebrate **stomach**, along with hydrochloric acid. The acid provides a suitable **pH** for pepsin, which digests long protein chains into shorter chains of **amino acids** called **peptides**.

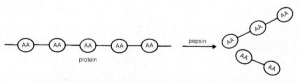

pepsin *The action of pepsin on protein.*

peptide A compound consisting of two or more **amino acids** linked between the amino group of one and the **acid** group of the next. The link between adjacent amino acids is called a *peptide bond*, and when many amino acids are joined in this way, the whole complex is called a *polypeptide*, which is the basis of **protein** structure.

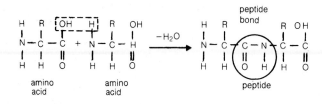

peptide *Two amino acids are joined in a peptide bond.*

peristalsis Waves of muscular contraction that pass along tubular organs and cause movement of their contents. For example, in mammals, peristalsis occurs in the **alimentary canal** and also in the **ureters** and **oviducts**. Peristalsis is caused by the rhythmic and coordinated contraction and relaxation of both circular and longitudinal **involuntary muscles**. *See also* **antagonistic muscles**.

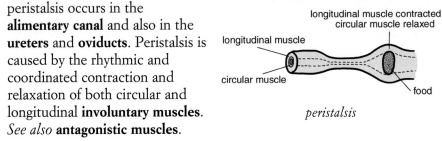

peristalsis

pest Any living organism that is considered to have a detrimental effect on humans.

Methods used to combat pests include:
(a) spraying with chemicals (**pesticides**);
(b) using natural **predators** against the pest;
(c) introducing **parasites** and **pathogens** to the pest population;
(d) introducing **sterile** individuals to the pest population, thus reducing reproductive capacity.

Example of pest	Effect on humans
Weeds, locusts	Reduce the growth of plants and crops
Foot and mouth virus	Causes disease in domestic animals
Woodworm, wet rot fungus	Damages buildings
Mosquitoes, lice	Transmit human disease

pesticide A chemical compound, often delivered in a spray, which kills pests or inhibits their growth. Examples:
(a) *herbicides*: weedkillers such as paraquat;
(b) *fungicides*: seed dressings such as organo-mercury compounds;
(c) *insecticides*: fly sprays used in the home; sprays released from aircraft against locusts. DDT (now banned in Britain) has been used against mosquitoes and lice.
The disadvantages of pesticides are:
(a) they may kill organisms other than the target pest;
(b) the concentration of a pesticide increases as it passes through a **food chain**;
(c) some decompose slowly and may accumulate into harmful doses within other organisms;
(d) by killing off susceptible organisms, they allow resistant individuals to grow and multiply with reduced **competition**.

pH A scale for measuring the acidity or alkalinity of a solution (*see* **acid**, **alkali**).

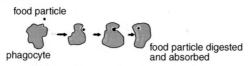

phagocytosis The process by which certain cells (*phagocytes*) surround and engulf a food particle, which is then digested. Phagocytosis is the feeding method employed by some **unicellular** protozoans, e.g. *Amoeba*. It is also one of the methods by which **white blood cells** destroy invading microorganisms.

food particle

phagocyte → food particle digested and absorbed

phagocytosis *The phagocyte engulfs and digests a food particle.*

pharynx A region of the vertebrate **alimentary canal** between the mouth and the **oesophagus**. In humans it is the back of the nose and throat that, when stimulated by food, initiates swallowing.

phenotype The physical characteristics of an organism resulting from the influence of **genotype** and environment. *See* **monohybrid inheritance**.

phloem Tissue within plants that transports carbohydrate from the leaves throughout the plant. Phloem consists of tubes that are formed from columns of living cells in which the horizontal cross-walls have become perforated. This allows the carbohydrate in aqueous solution to move from one phloem cell into the next and thus through the plant. Because of their structure, phloem tubes are also called *sieve tubes*. *See also* **companion cell**, **leaf**, **root**, **stem**, **translocation**.

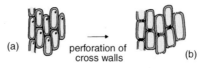

(a) perforation of cross walls (b)

phloem *Phloem cells (a) become sieve tubes (b).*

photoperiod The length of daylight that is important in many plant and animal **responses**, such as flowering in plants; migration in birds. *See also* **rhythmical behaviour**.

photoreceptor A sense organ that is stimulated by light, such as the eye.

photosynthesis The process by which green plants make **carbohydrate** from carbon dioxide and water. The energy for the reaction comes from sunlight, which is absorbed by the **chlorophyll** within **chloroplasts**. Oxygen is evolved as a by-product. Overall reaction:

$$\underset{6CO_2 + 6H_2O}{\underset{\text{dioxide}}{\text{carbon} + \text{water}}} \quad \xrightarrow[\text{chlorophyll}]{\text{light energy}} \quad \underset{C_6H_{12}O_6 \ + \ 6O_2}{\text{carbohydrate} + \text{oxygen}}$$

Photosynthesis is in fact a two-stage reaction involving:
(a) the *light reaction* in which light energy is used to split water into hydrogen (which passes to the next stage) and oxygen (which is released);
(b) the *dark reaction* in which the hydrogen from the light reaction combines with carbon dioxide to form carbohydrates.
It is the source of all food and the basis of **food chains**, while the release of oxygen replenishes the oxygen content of the atmosphere.

phototropism **Tropism** relative to light. Plant shoots are *positively phototropic*, i.e. they grow towards light.

light from one side →

phototropism

phylum A unit used in the **classification** of living organisms. A phylum consists of one or more **classes**. The term **division** is often substituted in plant classification.

phytoplankton *See* **plankton**.

pinna A flap of skin and **cartilage** at the outside end of the mammalian outer ear. *See* **ear**.

pitfall trap A trap used to collect organisms living on or just below the soil surface and from leaf litter. Examples of organisms caught are beetles and centipedes.

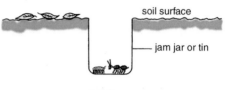

pitfall trap

pituitary gland An **endocrine gland** at the base of the vertebrate brain. It produces numerous hormones including **antidiuretic hormone** and **follicle-stimulating hormone**, many of which regulate the activity of other endocrine glands. The pituitary gland's own secretion is in many cases regulated by the brain.

placenta An organ that develops during **pregnancy** in the mammalian **uterus** and forms a close association between maternal and foetal blood circulations. The placenta allows the passage of food and oxygen to the **foetus** and removes carbon dioxide and **urea**.

plankton Microscopic animals (**zooplankton**) and plants (**phytoplankton**) that float in the surface waters of lakes and seas. Plankton are important as the basis of aquatic **food chains**.

plant growth substances Chemical compounds such as **auxins** that are involved in many plant processes including **tropisms**, **germination**, etc. The term 'plant hormone' was formerly used.

plasma The clear fluid of vertebrate **blood** in which the blood cells are suspended. It is an aqueous solution in which are dissolved many compounds that are in transit around the body. Among the compounds carried by plasma are: carbon dioxide and **urea** (both are waste products); **glucose** and **amino acids** (products of digestion); **hormones**; **plasma proteins**; and sodium chloride (salt).

plasma proteins **Proteins** dissolved in the **plasma** of vertebrate blood. Examples are **antibodies**, **fibrinogen**, and some **hormones**.

plasmid A circular group of **genes** found in bacterial cells, independent of the bacterial **chromosome**. *See also* **genetic engineering**.

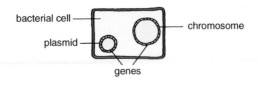

plasmid

plasmolysis The loss of water from a plant cell by **osmosis** when the cell is surrounded by a solution whose water concentration is less than that of the cell **vacuole** (for example, a strong sugar or salt solution). Osmosis causes water to pass out of the cell, making the vacuole shrink, and results in the pulling away of the **cytoplasm** from the cell wall.

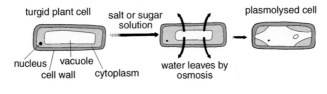

plasmolysis

Plasmolysis can be induced and reversed experimentally, but continued plasmolysis results in cell death. This osmotic death rarely occurs naturally but can result from adding excessive **fertilizer** to plants, which induces plasmolysis, or what is called 'plant burning'.

platelet The smallest cell found in mammalian blood. It is involved in **blood clotting**.

pleural membranes The two **membranes** that cover the outside of the **lungs** and line the inside of the **thorax** in mammals, and that secrete pleural fluid between them, so facilitating breathing movements.

plumule The leafy part of the embryonic **shoot** of seed plants. *See* **germination**.

pollen Reproductive cells of flowering plants, each containing a male **gamete**. Pollen grains are adapted to their mode of transfer, which may be either by insects or by wind. *See* figure opposite.

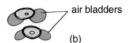

(a)

(b)

pollen (*a*) *Spiky and sticky for insect pollination.*
(*b*) *Smooth and light for wind pollination.*

pollination   The transfer of **pollen** grains from **stamens** to **carpels** in flowering plants. Pollination within the same flower or between flowers on the same plant is called *self-pollination*. Pollination between two separate plants is called *cross-pollination*. Normally male and female parts of the same plant do not mature simultaneously, favouring cross-pollination with a consequent mixing of **chromosomes**, which can lead to **variation**. Pollen is transferred on the bodies of insects or by the wind. Flowers are adapted to favour one particular method of transfer:

(a) *insect pollination*: insects visit flowers to drink or collect nectar, attracted by the colour or scent of the flower. Their bodies become dusted with pollen, some of which may adhere to the **stigmas** of subsequent flowers that they visit;

(b) *wind pollination*: pollen grains carried by the wind must be produced in much higher numbers than those carried by insects, to compensate for loss during transfer.

See also **fertilization**.

pollution   The addition of any substance to the environment, which upsets the natural balance. Pollution has resulted mainly from industrialization, which is largely based on the burning of **fossil fuels** and causes migration from the land to towns and cities.

(a) *Air pollution* is caused particularly by fossil fuel burning. Air pollutants such as smoke and sulphur dioxide cause irritation in the human respiratory system and may accelerate diseases such as bronchitis and lung cancer.

coal burning ⟶ smoke + carbon dioxide + sulphur dioxide

petrol burning ⟶ smoke + carbon monoxide + oxides of nitrogen +lead

(b) Water pollution results from the intentional or accidental addition of materials to either fresh water or sea water. The pollutants originate from industrial and agricultural practices and also from the home.

Examples are mine and quarry washings, acids, pesticides, radioactive wastes, fertilizers, detergents, oil, sewage and hot water (from power stations). Some pollutants, such as pesticides, may poison aquatic organisms, while organic pollutants, such as sewage, cause an increase in the microorganism population in the water. This decreases dissolved oxygen levels, making the water unfit for many organisms.

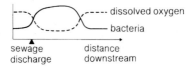

pollution *The effect of sewage discharge on oxygen levels in water.*

polyploidy A **mutation** in which an organism has extra sets of **chromosomes**. Polyploidy is more common in plants than in animals and often confers advantages such as increased disease resistance.

polysaccharides **Carbohydrates** consisting of long chains of **monosaccharides** that are linked together by chemical bonds. **Glucose** units can be linked in different ways to form several polysaccharides, such as **starch**, **glycogen** and **cellulose**.

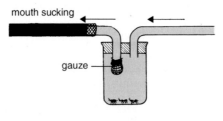

polysaccharides *A polysaccharide chain.*

pooter A device used to collect small insects by using suction.

pooter

population A group of organisms of the same **species** within a community.

posterior Relating to parts of the body at or near the hind end of an animal. *Compare* **anterior**.

predator An animal that feeds on other animals, which are called the **prey**, and which it catches; i.e. a predator is a **food consumer** but is not a **parasite**. The relationship between predator and prey can have dramatic effects on their numbers and on the food chain.

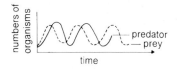

predator *A typical predator–prey relationship.*

pregnancy or gestation period The time from **conception** to **birth** in mammals. Human pregnancy lasts about 40 weeks, the **embryo** developing in the **uterus** after **implantation**. Finger-like structures (**villi**) grow from the embryo and develop into the **placenta**.

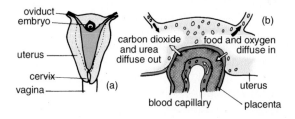

pregnancy *(a) Implanted embryo. (b) Relationship between uterus and placenta.*

During pregnancy the cells of the embryo continually divide and differentiate, and the growing embryo (**foetus**) becomes suspended in a water sac, the **amnion**. The placenta extends into the **umbilical cord**, which connects with the **abdomen** of the foetus.

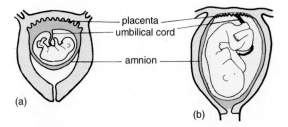

pregnancy *(a) Human foetus after twenty weeks.
(b) Human foetus just before birth.*

premolars **Teeth** that are found in front of **molars**.

prey An animal hunted for as food by another animal. *See* **predator**.

primary sexual characteristics Features that distinguish between males and females of the same species from the time of birth. They do not include those that develop at **puberty** and are characteristic of adulthood. *Compare* **secondary sexual characteristics**.

progeny The offspring of **reproduction**.

progesterone A **hormone** secreted by mammalian **ovaries**, which prepares the **uterus** for **implantation** and prevents further **ovulation** during **pregnancy**.

propagation *See* **vegetative reproduction**.

proprioreceptor A **receptor** that is stimulated by change in position of the body, e.g. stretch receptors in **muscle** fibres.

protease Any **enzyme** that breaks down **protein** into **peptides** or **amino acids**, by **hydrolysis**. Examples are **pepsin** and **trypsin**.

proteins **Organic compounds** containing the elements carbon, hydrogen, oxygen and nitrogen and consisting of long chains of subunits called **amino acids**. These chains may then be combined with others, and folded in several different ways, with various types of chemical bonding between chains and parts of chains, giving very large and complex molecules.

Proteins are the 'building blocks' of cells and tissues, being important constituents of muscle, skin, bone, etc. Proteins also play a vital role as enzymes while some hormones are protein in structure. *See also* **peptide**.

where (AA) = amino acid

proteins *Amino acid chains combine to form large and complex molecules.*

protein synthesis The synthesis of protein molecules from their constituent **amino acids**. The information for the construction of a protein with the correct sequence of amino acids is carried in the arrangement of nitrogen base pairs in **DNA**. This **genetic code** is transcribed exactly into **RNA**. The first stage is the unzipping of a DNA molecule.

Assume that the N-base sequence of the upper strand is to be transcribed. Nucleotide raw materials (*see* **nucleic acids**) bond to the appropriate N-base along the DNA strand.

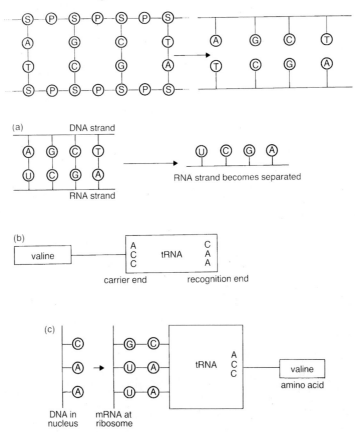

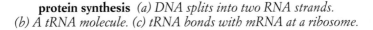

protein synthesis (*a*) *DNA splits into two RNA strands.*
(*b*) *A tRNA molecule.* (*c*) *tRNA bonds with mRNA at a ribosome.*

The finished RNA molecule separates from the DNA, which has acted as a template. The specific DNA code is now imprinted on the RNA molecule in the form of the corresponding bases. This code represents the information required for protein synthesis. The genetic code 'names' each amino acid by a sequence of 3 adjacent N-bases in DNA and then RNA. These *triplet codes* have been identified for the 20 to 24 amino acids found in living organisms.

The RNA containing the coded triplets is called **messenger RNA**. The mRNA molecules move towards and line up at **ribosomes**. Available in the

cytoplasm are the amino acids and another type of RNA called
transfer RNA.

Each tRNA molecule is a comparatively short polynucleotide at each end
of which is an important N-base triplet. At one end, the 'carrier' end, all
tRNA molecules have the same triplet: ACC, and this end links with an
amino acid. The other end, the 'recognition' end, has a triplet that is
specific for a particular amino acid, e.g. the amino acid valine has the
specific triplet CAA.

When the tRNA arrives at a ribosome, the specific triplet will be able to
bond only to a corresponding triplet along the mRNA. In this way, amino
acids become positioned along the mRNA in a code-determined sequence.
By the formation of **peptide** bonds between adjacent amino acids,
polypeptides and hence proteins are synthesized.

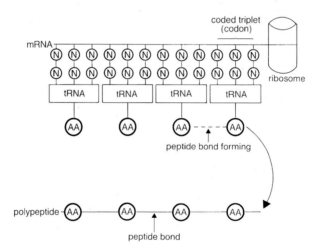

protein synthesis *Amino acids bond at a ribosome to
form a polypeptide.*

Protista **Unicellular** organisms with varied types of nutrition, including
photosynthesis. Reproduction may be sexual or asexual. **Cilia** or **flagella**
may be present. *See also* **Protozoa**.

proton A negatively charged particle that occurs in the nucleus of
all **atoms**.

protoplasm All the material within and including the cell membrane (*see*
cell), i.e. protoplasm consists of a **nucleus** and **cytoplasm**.

protozoa A **phylum** consisting of **unicellular** animals such as *Amoeba* and *Paramecium*. Protozoans live in a wide variety of habitats including stagnant water and faeces. They feed on bacteria and some cause disease in humans, such as dysentery and malaria.

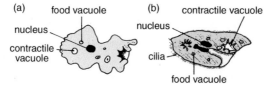

protozoa *(a) Amoeba (b) Paramecium.*

puberty The period in the human **life cycle** when a person becomes sexually mature. In females, usually at the age of 10–14 years, **ovulation** begins, and in males at about the same age production of **spermatozoa** starts. The **secondary sexual characteristics** develop in both sexes during puberty.

pulmonary Related to the **lungs** and **breathing**. The term is applied to organs, tissues and parts of the body.*See also* **pulmonary vessels**, **heart**, **circulatory system**.

pulmonary vein *See* **pulmonary vessels**, **heart**, **circulatory system**.

pulmonary vessels Mammalian **blood vessels** that because of their special functions, do not obey the general rule that **arteries** carry oxygenated blood, and **veins** carry deoxygenated blood. The *pulmonary artery* carries deoxygenated blood from the right **ventricle** to the **lungs**, and the *pulmonary vein* carries oxygenated blood from the lungs to the left **atrium**. *See also* **heart**, **circulatory system**.

pulse rate The regular beating in **arteries** due to the rhythmic movement of **blood** resulting from the **heartbeat**. Pulse rate can be detected in the human body where an artery is close to the skin surface, for example, at the wrist. In an adult human, pulse rate varies from about 70 beats per minute at rest, to over 100 beats per minute during exercise.

Punnett square A graphic method used in **genetics** to calculate the results of all possible **fertilizations** and hence the **genotypes** and **phenotypes** of **progeny**. In a Punnett square, the symbols used to represent one of the parent's **gamete** genotypes are written along the top and those of the other parent down the side. The permutations possible during fertilization are worked out by matching male and female gametes. *See also* **backcross, incomplete dominance, monohybrid inheritance.**

pupa A stage in the **life history** of some insects between **larva** and **imago**, during which a radical change in form occurs. *See also* **metamorphosis.**

pure-breeding Possessing an inherited trait, controlled by a **homozygous** pair of **alleles**, which in successive self-crosses reappears generation after generation. *See also* **monohybrid inheritance.**

pyloric sphincter The ring of muscles that opens and closes the entry to the **stomach.**

pyramid of biomass A diagram illustrating that organisms at the end of a **food chain** have a smaller total mass (**biomass**) than those at the beginning of the chain. *See also* **pyramid of numbers.**

pyramid of numbers A diagram illustrating the relationship between members of a **food chain**, showing that the organisms at the end of the chain are usually fewer in number, or to be more exact, have a smaller total mass (**biomass**) than those at the beginning of the chain.

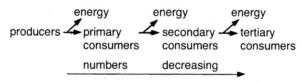

pyramid of numbers *Energy loss.*

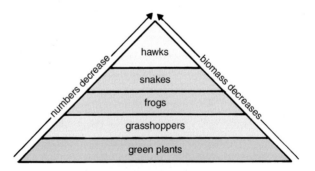

pyramid of numbers

The decrease is caused by **energy** losses at each link in the chain, i.e. each organism in a food chain uses up energy in various activities such as heat production and movement. This energy is lost to the subsequent organisms in the chain and so the reduced energy can only support a smaller number of individuals.

quadrat A square frame used to sample plants (and stationary animals, e.g. barnacles). A quadrat marks off a small area so that the **species** present can be identified and counted. This sample gives an estimate of the numbers and types of species in the whole area. The quadrat must be placed randomly to obtain a representative sample and a number of samples taken.

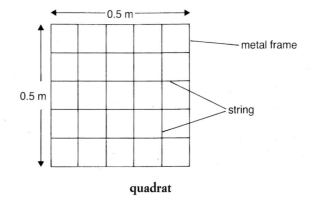

quadrat

R

radicle The embryonic root of seed plants. It is the first structure to emerge from the seed during **germination**.

radius The **anterior** of the two bones of the lower region of the **tetrapod** forelimb. In humans, it is the shorter of the two bones of the forearm. *See* **endoskeleton**.

receptor or sense organ A specialized animal tissue animal that detects stimuli from the environment and which, by sending **nerve impulses** through the **nervous system**, causes **responses** to be made. *See* **sensitivity**.

recessive One of a pair of **alleles** that is only expressed in a **homozygous phenotype**. It is the converse of **dominant**. *See also* **backcross, incomplete dominance, monohybrid inheritance**.

rectum The terminal part of the vertebrate **intestine** in which **faeces** are stored prior to expulsion via the **anus** or **cloaca**. *See* **digestion**.

red blood cell, red blood corpuscle or erythrocyte The most numerous cell of vertebrate **blood**, responsible for transporting oxygen from the lungs to the tissues. In humans, red blood cells are made in bone marrow and are biconcave discs, without **nuclei**.

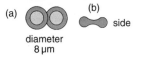

(a) diameter 8 μm   (b) side

red blood cell *Shape as seen (a) from above and (b) from the side.*

Red blood cells contain **haemoglobin**, which combines with oxygen as blood passes through the lungs, forming a compound called *oxyhaemoglobin*. At the tissues, this unstable compound breaks down, thus releasing oxygen to the cells.

A shortage of red blood cells is called *anaemia*.

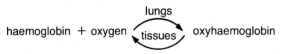

lungs

haemoglobin + oxygen ⇄ tissues ⇄ oxyhaemoglobin

red blood cell *Haemoglobin delivers oxygen to the tissues.*

red blood corpuscle *See* **red blood cell**.

reflex action A rapid involuntary **response** to a **stimulus**, occurring in most animals and in vertebrates, mediated by the **spinal cord**. Reflex actions can be important in protecting animals from injury, by for instance withdrawing a limb from a hot object.

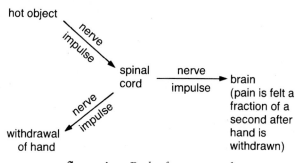

reflex action *Path of nerve impulses.*

The structures involved and the **nerve impulses** responsible for reflex actions constitute a *reflex arc* which is set up when a nerve impulse is initiated at a receptor. The impulse is transmitted along a sensory **neurone** to the spinal cord where it crosses a **synapse** to a motor neurone.

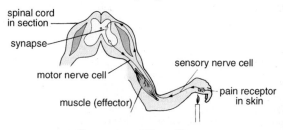

reflex action *The reflex arc.*

When a reflex arc operates, nerve impulses are also sent from the spinal cord to the brain. Thus, although the response is initiated by the spinal cord, it is through the brain that the animal is aware of what has happened.

regeneration The regrowth by an organism of tissues or organs that have been damaged or removed. In certain lower animals whole new individuals can develop from portions of a damaged adult. For example, a starfish may develop a new arm if one is removed, and the amputated portion may grow a new body.

Regeneration is common among plants. This fact is used by gardeners who grow plants from cuttings. *See also* **artificial propagation**.

In higher animals, this degree of regeneration is not possible, due to the complexity of the cells and tissues present in such animals. Thus in mammals, wound healing is the only form of regeneration possible.

renal Related to the **kidneys** and **urine** production.

reproduction The process by which a new organism is produced from one or a pair of parent organisms. *See* **asexual reproduction, sexual reproduction**.

respiration The reactions by which organisms release the chemical energy contained in food, such as **glucose**. The energy is used to synthesize **ATP** from ADP and is then available for other metabolic processes, for example, muscle action. There are two kinds of respiration:
(a) *aerobic respiration*: this occurs in the presence of oxygen within the **mitochondria** of cells;

glucose + oxygen $\longrightarrow$ carbon + water dioxide

$C_6H_{12}O_6 + 6O_2$ ADP   ATP $6CO_2 + 6H_2O$

respiration *Aerobic respiration.*

(b) *anaerobic respiration*: this occurs in the absence of oxygen within the **cytoplasm** of cells, and provides a lower ATP yield than aerobic respiration.

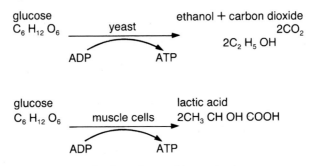

glucose
$C_6H_{12}O_6$ ——— yeast ——→ ethanol + carbon dioxide
                                    $2CO_2$
                              $2C_2H_5OH$
ADP   ATP

glucose
$C_6H_{12}O_6$ ——— muscle cells ——→ lactic acid
                                    $2CH_3\,CH\,OH\,COOH$
ADP   ATP

respiration *Anaerobic respiration.*

response Any change in an organism made in reaction to a **stimulus**. *See* **sensitivity**.

retina Light-sensitive tissue lining the interior of the vertebrate **eye**, and consisting of two types of cell: **rods** and **cones**.

rhizome An organ of **vegetative reproduction** in flowering plants. It consists of a horizontal underground stem growing from a parent plant. The tip of the rhizome

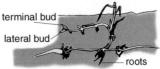

terminal bud
lateral bud
roots

rhizome *A grass rhizome.*

is a bud from which grows a new plant. Plants that have rhizomes include the iris and many types of grass.

rhythmical behaviour Animal behaviour patterns that are repeated at definite time intervals, e.g. once per day or once per year. Such behaviour is normally triggered by an external stimulus such as changes in daylength, and are controlled internally by the animal's **biological clock**. Examples are courtship behaviour, migration and hibernation.

ribosomes Microscopic cell organelles in the **cytoplasm**. They are the sites of **protein synthesis**.

RNA (ribonucleic acid) **Nucleic acid** synthesized by **DNA** in the **nucleus** of cells. It is responsible for carrying the **genetic code** from the nucleus into the **cytoplasm** where **protein synthesis** occurs.
  RNA differs from DNA in the following ways:
(a) RNA is a single polynucleotide chain;
(b) the sugar group is ribose;
(c) thymine is replaced by uracil.

rod A light-sensitive **neurone** in the **retina** of the vertebrate **eye**, which can function in dim light.

root The part of a flowering plant that normally grows down into the soil. Its functions are:
(a) absorption of water and **mineral salts** from soil;
(b) the anchoring of the plant in the soil;
(c) the storage of food, in some plants, e.g. turnip.

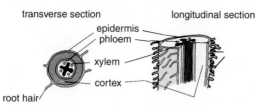

transverse section   longitudinal section

epidermis
phloem
xylem
cortex
root hair

root *The structure of a dicotyledon root.*

root cap A cap-shaped layer of cells, covering the apex of the growing **root** tip and protecting it as the root grows through the soil.

root tip — — root cap

root cap *Longitudinal section through a root.*

root hairs Tubular projections from root **epidermis** cells, the **nucleus** usually passing into the hair. Root hairs enormously increase the surface area of the root, and are the principal absorbing tissue of the plant. Water enters root hairs from the soil by osmosis, while **mineral salts** are absorbed by **active transport**. The water and mineral salts then pass through the **cortex** cells and enter **xylem** vessels from where they are transported throughout the plant via the **transpiration stream**.

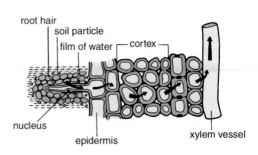

root hairs *Water and mineral salts enter the plant via the root hairs.*

root nodules Swellings on the roots of leguminous plants (for example, clover, bean, pea). Root nodules contain bacteria of the **genus** *Rhizobium* that convert the nitrogen of soil-air into organic nitrogen compounds, which can be used by the legumes. This is called **nitrogen fixation**. *See also* **nitrogen cycle**.

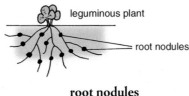

root nodules

roughage An important component of the human **balanced diet**, consisting mainly of the **cellulose** in plant cell walls. Although indigestible by human beings, roughage is important in the diet as it adds bulk to food and enables the muscles of the **alimentary canal** to grip the food and keep it moving by **peristalsis**.

S

saliva Fluid secreted by *salivary glands* into the mouths of many animals in order to moisten and lubricate food. In some mammals, including humans, saliva contains the enzyme *salivary amylase*.

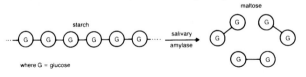

saliva *Amylase in saliva digests starch into maltose.*

saprophyte An organism that feeds on dead and decaying plants and animals, causing decomposition. Many **fungi** and **bacteria** are saprophytic and play an important role in recycling nutrients. *See also* **carbon cycle**, **nitrogen cycle**.

scapula The **dorsal** part of the **tetrapod** shoulder girdle. In humans, it is the shoulder blade. *See* **endoskeleton**.

scientific method The procedures by which scientific investigations should be made. Scientific method involves the following steps:
(a) observation: an occurrence is seen to happen on more than one occasion. For example, it may be observed that **starch** in plant seeds apparently supplies energy during **germination**;
(b) problem identification: the observation is questioned. For example, how does starch, which is a long chain carbohydrate become suitable as a respiratory **substrate**?
(c) **hypothesis**: the suggestion of a possible solution. For example, a **carbohydrase** enzyme within seeds degrades starch to **glucose**;
(d) Experiment: the hypothesis is tested. For, example, seed extract is added to starch, and a test for glucose is made;
(e) **theory**: the proposal of a solution to the problem based on experimental evidence; for example, that in plant seeds, a carbohydrase enzyme degrades starch to glucose, which then acts as a respiratory substrate during germination.
All valid scientific investigations follow the guidelines of the scientific method and must include **control experiments**, which are identical to the test experiment in all aspects except one. The control provides a standard with which the test experiment can be compared, by showing that any change occurring in the test experiment was due to the factor missing from

the control and would not have happened anyway. For example, in the seed/starch experiment, a tube containing starch alone would be a suitable control.

sclerotic The external protective layer of the vertebrate eyeball. *See* **eye**.

secondary sexual characteristics Features (excluding **gonads** and associated structures) that distinguish between adult male and female animals. The development of such features, for example, lion's mane, stag's antlers, and in humans, breast development in females, facial hair in males, etc., is usually controlled by sex **hormones**. *Compare* **primary sexual characteristics**.

seed The structure that develops from an **ovule** after fertilization in flowering plants, and which grows into a new plant. Seeds are enclosed within a **fruit**.

Within the seed, the **embryo** becomes differentiated into an embryonic **shoot** bud (**plumule**) and **root** (**radicle**) and either one or two seed leaves (**cotyledons**).

See also **fruit and seed dispersal, germination**.

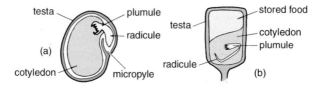

seed *The structure of a seed: (a) broad bean (b) maize grain.*

selective breeding *See* **artificial selection**.

selectively permeable membrane Membranes that surround cells and the bordering regions within cells. They are made up of orderly arrangements of protein and fat molecules. Certain small molecules may pass through pores in the membrane but larger ones are held back. For this reason the membrane is described as selectively permeable or *semipermeable*. *See also* **cell**, **osmosis**.

semicircular canals Tubes within the vertebrate inner ear, which are important in maintaining balance. *See* **ear**.

sense organ *See* **receptor, sensitivity**.

sensitivity or irritability The ability of living organisms to respond to changes in environmental **stimuli**, such as heat, light, sound, etc. Sensitivity enables organisms to be aware of changes in their environment and thus to make appropriate **responses** to any changes that may occur. Certain parts of animals, for example, eyes, ears, skin, are sensitive to particular environmental stimuli and are called sense organs or **receptors**. Similarly, plant tissues such as shoot tips, are receptors, being important in **tropisms**.

Sense	Stimulus	Receptor
smell	chemicals	nose
taste	chemicals	mouth
touch	contact	skin
hearing	sound	ears
sight	light	eyes
balance	change of position	inner ear

As a result of stimuli from the environment, responses are initiated in specialized structures called **effectors**, for example, muscles. The responses made by an organism constitute its behaviour.

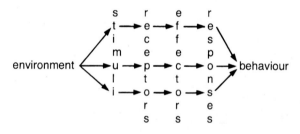

sensitivity *Organisms respond to changes in their environment.*

In mammals the receptors are specialized cells connected to the brain or spinal cord, which together make up the **central nervous system** (CNS). In response to a stimulus, the receptors initiate a **nerve impulse**, which is transmitted by **neurones** to the CNS, i.e. the receptor converts the energy of the stimulus into the electrical energy of the nerve impulse. *See also* **smell**, **taste**.

sewage disposal The treatment of sewage in order to make it harmless. Sewage, which is mainly domestic waste, is first filtered to remove large particles, and then piped into tanks where solids settle out as *sludge*. The remaining liquid is aerated to encourage the growth of *aerobic bacteria*, which feed on dissolved organic compounds. The bacteria are, in turn, eaten by larger organisms and when the dead organisms are allowed to settle, the remaining liquid can be safely discharged into a river. Sludge may also be digested by **microbe** action, with the production of **methane** gas (also called *biogas*), which is used as a fuel.

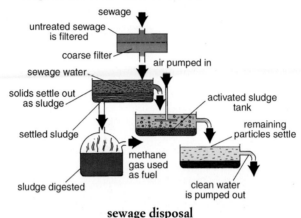

sewage disposal

sex chromosome Any chromosome that is involved in **sex determination**. In **diploid** human cells there are 46 chromosomes made up of 23 **homologous chromosome** pairs, one of the pairs being described as sex chromosomes. In the female, the two sex chromosomes are similar (*homogametic*) and are called X chromosomes. The female **genotype** is thus XX. In the male, one of the pair is distinctly smaller, and is called the Y chromosome. The male genotype is thus XY (*heterogametic*).

The male is not always the heterogametic sex. For example, in birds, the male is XX, and the female is XY, while in some insects the female is XX and the male is XO, the Y chromosome being absent.

The sex chromosomes, as well as determining sex, also contain **genes** controlling other traits, resulting in what is known as **sex linkage**.

sex determination The method by which the sex of a **zygote** is determined, the most common method being by **sex chromosomes**. Consider the **genotypes** of (a) a human male and (b) a human female:
(a) a Y-bearing sperm may fertilize an ovum giving a zygote genotype XY and **phenotype** male;
(b) an X-bearing sperm may fertilize an ovum giving a zygote genotype XX and phenotype female. Since half the sperms are X and half are Y, there is an equal chance of the zygote being male or female.

sex linkage The presence on a **sex chromosome** of **genes** unconnected with sexuality. It results in the appearance of certain traits in one sex only. In humans, most sex-linked genes are carried on X chromosomes, the Y chromosome being concerned mainly with sexuality.

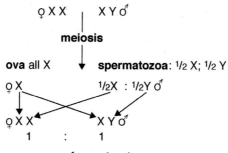

sex determination

As an example, consider colour blindness in humans: The gene for colour blindness is carried on the X chromosome. Normal vision is **dominant** to colour blindness.

If N = normal and n = colour blind:

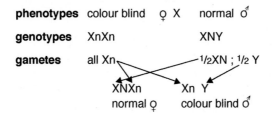

sex linkage *Colour blindness.*

In this case, the **heterozygous** / XNXn is called a 'carrier' since she has normal vision but carries the **recessive allele**. Thus if crossed with a normal male:

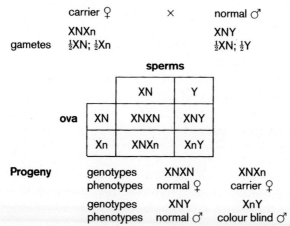

	carrier ♀	×	normal ♂
	XNXn		XNY
gametes	½XN; ½Xn		½XN; ½Y

sperms

		XN	Y
ova	XN	XNXN	XNY
	Xn	XNXn	XnY

Progeny	genotypes	XNXN	XNXn
	phenotypes	normal ♀	carrier ♀
	genotypes	XNY	XnY
	phenotypes	normal ♂	colour blind ♂

sex linkage *The inheritance of colour blindness from a carrier.*

The result from the **Punnett square** indicates a possibility that half the sons will be colour blind, and half the daughters will be carriers.

A more serious sex-linked trait is **haemophilia** (prolonged bleeding) but its transmission and inheritance are the same as above.

sexual reproduction **Reproduction** involving the joining or fusing of two sex cells (**gametes**), one from a male parent and one from a female parent. Gametes are **haploid** and when they fuse (**fertilization**), the resulting composite cell (**zygote**) has the **diploid** number of **chromosomes**. After fertilization, the diploid zygote divides repeatedly, ultimately resulting in a new organism. The diagram shows this for humans, who have 46 chromosomes.

sexual reproduction *Two haploid gametes fuse to form a diploid zygote.*

The offspring of sexual reproduction are genetically unique (except for identical twins) because, unlike the offspring of **asexual reproduction**, they obtain half their chromosomes from their male parent and half from their female parent. Thus each fertilization produces a new combination of chromosomes, which in turn produce a new organism that will likewise produce gametes by **meiosis**.

shoot The part of a flowering plant that is above soil level, for example, stem, leaves, buds and flowers.

short sight **or** **myopia** A human eye defect, mainly caused by the distance from the **lens** to the **retina** being longer than normal. This results in distant objects being focused in front of the retina giving blurred vision. Short sight is corrected by wearing diverging (*concave*) lenses.

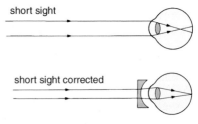

short sight

single-cell protein (SCP) A protein-rich food supplement produced by certain microorganisms, which is fed to domestic animals. Since microorganisms reproduce very rapidly, and are more efficient protein producers than other organisms, they can be grown, harvested and dried to produce SCP.

The figures below compare the protein produced per day by 1000 kg of **biomass**:

cattle 1 kg
soya beans 100 kg
microorganisms 10^{15} kg.

skeleton The hard framework of an animal, which supports and protects the internal organs. *See* **endoskeleton, exoskeleton.**

skin The layer of epithelial cells (*see* **epithelium**), **connective tissue** and associated structures that covers most of the body of vertebrates. Mammalian skin consists of two main layers:

(a) the **epidermis**; this is the outer layer. It consists of
  (i) *cornified layer*: dead cells forming a tough protective outer coat;
  (ii) *granular layer*: living cells, which ultimately form the cornified layer;
  (iii) *Malpighian layer*: actively dividing cells, which produce new epidermis;
(b) the **dermis**; this is a thicker layer below the epidermis, containing blood capillaries, hair follicles, sweat glands and **receptor** cells sensitive to touch, heat, cold, pain and pressure.

Beneath the dermis, there is a layer of fat storage cells, which also act as heat insulation.

The functions of mammalian skin are:
(a) to protect against injury and microorganism entry;
(b) to reduce water loss by evaporation;
(c) to act as **receptors** for certain environmental stimuli;
(d) to play an important part in **temperature regulation** in **endotherms**.

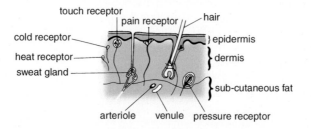

skin *Section through mammalian skin.*

small intestine The **anterior** region of the vertebrate **intestine**. In man, it consists of the **duodenum** (about thirty centimetres in length) and the **ileum** (about seven metres in length). The duodenum receives food from the **stomach**. *See* **digestion**.

smell The ability of animals to detect odours. In humans, the **receptor** cells involved are in the nasal cavity, and are sensitive to chemical **stimuli**. *See* **sensitivity**.

smog A harmful side-effect of smoke **pollution** caused by particles of smoke sticking to droplets of water in the atmosphere forming a thick mist:

$$\text{smoke} + \text{fog} \rightarrow \text{smog}$$

Such a smog is thought to have resulted in 4000 deaths from respiratory disease in London in 1952. Since the Clean Air Act of 1956, smog has largely been eliminated in Britain, but is still prevalent in other industrialized countries.

smoking The drawing into the lungs of tobacco smoke, usually from cigarettes. Smoking is habit-forming and can cause lung cancer, bronchitis and other serious diseases.

smooth muscle *See* **involuntary muscle**.

soil The weathered layer of the Earth's crust intermingled with living organisms and the products of their decay.
   The components of soil are:
(a) inorganic particles (weathered rock fragments);
(b) water;
(c) **humus**;
(d) air;
(e) **mineral salts**;
(f) microorganisms;
(g) other organisms (for example, earthworms).
   Soil is important because:
(a) it is a **habitat** for a wide variety of organisms;
(b) it provides plants with water and mineral salts;
(c) decomposition of dead organisms in soil releases minerals that can be
   used by other living organisms.

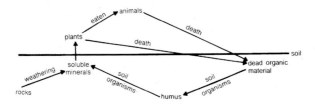

soil *Dead organic material is broken down in the soil.*

soil depletion The loss of **mineral salts** from **soil** when a crop is harvested, which like **soil erosion** may render the soil infertile. Soil depletion can be prevented by:
(a) **crop rotation**;
(b) the addition of **fertilizers**.

soil erosion The loss of soil due to the agricultural practices associated with crop growing, such as repeated ploughing, deforestation, etc., which make the top-soil, which is rich in **humus** and minerals, less stable and more vulnerable to the effects of wind and rain. *See also* **soil depletion**.

soil sieve A device used to separate the four types of inorganic particles in soil. The proportion of these particles in soil can be measured by passing a weighed sample of dried soil through sieves of varying mesh size, which separate the particles by size. The separated particles are weighed and their percentage of the complete sample can be calculated. *See also* **soil texture**.

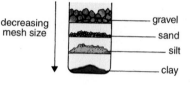

soil sieve

soil texture The types and proportions of inorganic particles in **soil**, of which four types are recognised, based on size.
   Soil texture has important effects on soil properties such as water retention and aeration. *See* **soil types**.

soil texture

soil thermometer A Celsius thermometer adapted for measuring **soil** temperature.

soil thermometer

soil types The numerous types of **soil** that exist. A simple classification recognizes three distinct types:

(a) sandy (light) soil has a high proportion of the larger inorganic particles, and hence larger air spaces. Thus, sandy soil is well aerated and has good drainage but tends to lose **mineral salts**, which are washed downwards (**leaching**);

(b) clay (heavy) soil has a high proportion of small particles, which means that it retains minerals, but is poorly aerated and can become waterlogged;

(c) loam soil is the most fertile soil, consisting of a balance of particle types and a good **humus** content. Soils of this type are well-aerated and drain freely, but still retain water and minerals.

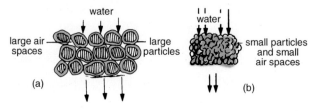

soil types *(a) Sandy soil. (b) Clay soil.*

solute The substance, in a **solution**, which dissolves in the **solvent**.

solution The mixture (usually a liquid) formed when one substance (the **solute**) dissolves in another (the **solvent**),

$$\text{solute} + \text{solvent} \rightarrow \text{solution}$$

e.g. sugar + water → sugar solution.

solvent The substance, in a **solution**, in which a **solute** dissolves.

species A unit used in the **classification** of living organisms. It is a group of organisms that share the same general physical characteristics and which can mate and produce fertile offspring. For example, all dogs, despite variation in shape, size etc., are of the same species, but horses and donkeys are separate species within the same **genus**.

spermatozoon (*pl.* **spermatazoa**) or **sperm** The small motile male **gamete** formed in animal **testes**, and usually having a **flagellum**. Sperms are released from the male in order to fertilize the female gamete. *See also* **fertilization**, **meiosis**.

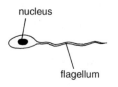

spermatazoon

sphincter A ring of muscle around tubular organs, which by contracting, can narrow or close the passage within the organ. Examples are the anal sphincter (at the **anus**) and the **pyloric sphincter**.

spinal cord That part of the vertebrate **central nervous system** that is enclosed within and protected by the **vertical column** (backbone).

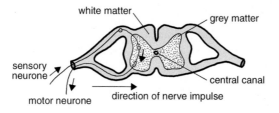

spinal cord *Section showing the three regions.*

The spinal cord is a cylindrical mass of **neurones** that connect with the **brain** and also with other parts of the body via *spinal nerves*. The spinal cord consists of three regions:
(a) an inner layer of grey matter consisting of neuronal cell bodies,
(b) an outer layer of white matter consisting of nerve fibres running the length of the cord;
(c) a fluid-filled central canal.
The spinal cord conducts **nerve impulses** to and from the brain and is also involved in **reflex actions**.

spiracle One of many pores in the **cuticle** of insects, connecting the **tracheae** with the atmosphere. *See* **gas exchange**.

spleen An abdominal organ (*see* **abdomen**), near the stomach, in most vertebrates. It produces **white blood cells**, destroys worn out **red blood cells** and filters foreign bodies from the blood.

spongy mesophyll Tissue in a leaf situated between the **palisade mesophyll** and the lower **epidermis**. Spongy mesophyll cells are loosely packed, being separated by air spaces, which allow **gas exchange** between the leaf and the atmosphere via the **stomata**.

spore A reproductive unit, usually microscopic, consisting of one or several cells, which becomes detached from a parent organism and

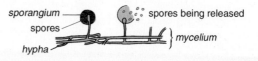

spore *Spore release in the bread mould Mucor.*

ultimately gives rise to a new individual. Spores are involved in both **asexual reproduction** and **sexual reproduction** (as **gametes**) and are produced by certain plants, fungi, bacteria and **protozoa**. Some spores form a resistant resting stage of a **life history** while others allow rapid colonization of new **habitats**.

stamen The male part of a **flower** in which pollen grains are produced. Each stamen consists of a stalk (*filament*) bearing an **anther**.

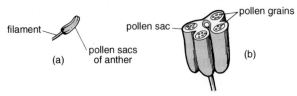

stamen (*a*) *Stamen.* (*b*) *Anther cut open.*

starch A **polysaccharide** carbohydrate that consists of chains of **glucose** units and which is important as an energy store in plants. Starch is synthesized during **photosynthesis** and is readily converted to glucose by **carbohydrase** enzymes. *See* **polysaccharides**.

stem That part of a flowering plant that bears the buds, leaves and flowers. Its functions are:
(a) the transport of water, **mineral salts** and carbohydrate;
(b) the raising of the leaves above the soil for maximum air and light;
(c) the raising of the flowers, thus aiding **pollination**;
(d) (in green stems) photosynthesis.

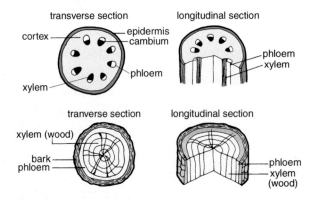

stem *Structure of the stem of a dicotyledonous plant. Transverse and longitudinal sections of (a) a young stem and (b) an older stem.*

sterilization **1.** A procedure to make an organism incapable of **reproduction**.

2. A procedure to make materials free of microorganisms. *See also* **autoclave**.

sternum (or **breastbone**) A bone in the middle of the **ventral** side of the **thorax** of **tetrapods**, to which most of the ventral ribs are attached. *See* **endoskeleton**.

stigma A sticky structure in a flower that traps incoming pollen during **pollination**. *See* **fertilization**.

stimulus Any change in the **environment** of an organism, which may provoke a **response** in the organism. *See* **sensitivity**.

stolon An organ of **vegetative reproduction** in flowering plants. It consists of a horizontal stem growing from a bud on the parent organism's stem. Stolons grow above the soil and eventually the tip becomes established in the soil and develops into an independent plant.

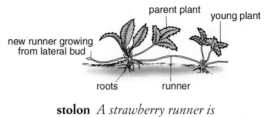

stolon *A strawberry runner is an example of a stolon.*

stoma (*pl.* **stomata**) One of many small pores in the **epidermis** of plants, particularly leaves. The evaporation of water during **transpiration** and **gas exchange**, occur via the stomata. *See also* **guard cells**.

stomach The muscular sac in the **anterior** region of the **alimentary canal**. In vertebrates, food is passed to the stomach by **peristalsis** via the **oesophagus**. In the stomach, food is mechanically churned by the peristaltic action of the walls, and protein digestion is initiated (*see* **pepsin**). From the stomach, food is passed into the **small intestine** through the **pyloric sphincter**.
In **herbivores**, the stomach has several chambers for **cellulose** digestion.

striated muscle *See* **voluntary muscle**.

substrate A substance that is acted upon by an **enzyme**.

sugars Water-soluble, sweet-tasting crystalline **carbohydrates**, which include the **monosaccharides** and the **disaccharides**.

surface area/volume ratio The important relationship between the surface area of an organism or structure (such as a cell) and its overall volume or mass, which is significant to living organisms in several ways.

It is difficult to measure the surface area and volume of a plant or animal, but by using cubes as model organisms, the importance of the ratio can be seen. As the object becomes larger, its surface area becomes smaller relative to its volume. In living organisms this ratio has special significance in terms of heat and water loss.

$$\frac{\text{surface area}}{\text{volume}} \quad \text{or} \quad \frac{\text{surface area}}{\text{mass}}$$

(a) Surface area/volume and heat loss: heat is lost more rapidly from small animals across their relatively large surface area, with the following consequences:
(i) small mammals such as mice eat relatively more food than larger mammals in order to generate energy to replace their high heat losses;
(ii) very small birds and mammals are restricted to warm climates;
(iii) birds and mammals in cold **habitats** are usually larger than the same species living in warm climates.

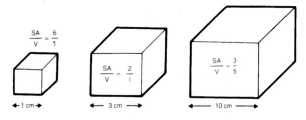

surface area/volume ratio

(b) Surface area/volume and water loss: relative to their volume, small organisms have a larger evaporating surface and thus a greater tendency to lose water. This is important because many animals and plants have problems controlling water balance, so the smaller the organism the greater the problem.

suspensory ligaments Structures holding the **lens** in place in the vertebrate **eye**. *See also* **accommodation**.

symbiosis A relationship between organisms of different species for the purpose of nutrition. Examples of symbiosis include **parasitism**, **mutualism** and **commensalism**, although the term is sometimes restricted to mutualism.

synapse A microscopic gap between the **axon** of one **neurone** and the *dendrites* of another, across which a **nerve impulse** must pass. Nerve impulses arriving at a synapse cause **diffusion** of a chemical substance, which crosses the gap to initiate nerve impulses in the next neurone.

nerve impulse	synapse	nerve impulse
axon	diffusion of chemical	dendrite

synapse *A chemical substance diffuses across the synapse and initiates nerve impulses in the next nerve cell.*

synovial membrane The membrane of **connective tissue** lining the capsule of a vertebrate moveable joint, being attached to the bones at either side of the joint.

The synovial membrane secretes *synovial fluid*, which bathes the joint cavity, lubricating the joint when the bones move, and cushioning against jarring.

systole *See* **heartbeat.**

T

taste The ability of animals to detect flavours. In humans, the **receptor** cells involved are *taste buds*, which are sensitive to chemical **stimuli**, and are restricted to the mouth, particularly the tongue. There are four types of taste bud, sensitive to sweetness, sourness, saltiness and bitterness. *See* **sensitivity**.

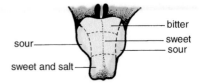

taste *A taste-map of the tongue.*

taxis (*pl.* **taxes**) A locomotory movement of a simple organism or a cell in response to an environmental stimulus such as light.

Such movements show a relationship to the direction of the stimulus, the movement either being toward (positive) or away (negative) from the source of the stimulus.

Taxes are named by adding a prefix that relates to the stimulus. Thus a taxis relative to light is a *phototaxis*. Some examples of taxes:
(a) Paramecium is negatively *geotactic*, i.e. it swims away from gravity;
(b) Fruit flies are positively *phototactic*, i.e. they move toward light;
(c) Many **spermatozoa** are positively *chemotactic*, i.e. they move toward chemical substances released by **ova**.

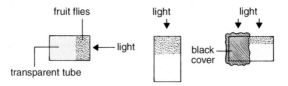

taxis *Positive phototaxis in fruit flies.*

teeth Structures within the vertebrate mouth that are used for biting, tearing and crushing food before it is swallowed. Teeth consist of the following materials:
(a) *enamel*: a hard substance covering the exposed surface of the tooth (the *crown*). It contains calcium phosphate and provides an efficient biting surface.
(b) *dentine*: a substance similar to bone, forming the inner part of the tooth.
(c) *pulp*: soft tissue in the centre of the tooth containing blood capillaries,

which supply food and oxygen, and sensory **neurones**, which register pain if the tooth is damaged.

(d) *root*: the part of the tooth within the gum, which is embedded in the jawbone by a substance called *cement*.

The types of teeth are **canines, carnassials, incisors, molars** and **premolars**. *See also* **carnivore, dental formula, dentition, herbivore, omnivore**.

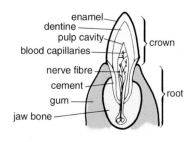

teeth *Structure of a tooth.*

temperature regulation The mechanisms, in **endotherms**, involved in maintaining body temperature within a narrow range (for example, in humans, close to 37 °C) so that the normal reactions of **metabolism** can take place.

Some of the temperature regulators employed by birds and mammals are outlined below:

(a) fat under the skin (*subcutaneous fat*) acts as an insulator;

(b) hair in mammals, and feathers in birds, trap air, which is a good insulator;

(c) evaporation of sweat from the skin surface of some mammals has a cooling effect;

(d) *vasoconstriction*: superficial blood vessels constrict in response to cold, diverting blood away from the skin surface, and thus reducing heat loss;

(e) *vasodilation*: superficial blood vessels dilate in response to heat, bringing blood to the skin surface, from which heat can be lost to the atmosphere.

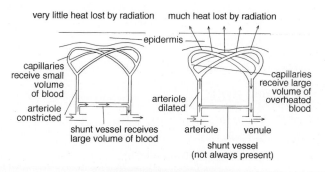

temperature regulation
Vasoconstriction and vasodilation.

tendon A band of **connective tissue** by which muscles are attached to bones.

tensile strength The ability to withstand a certain amount of bending before breaking. *See* **bone, collagen**.

testa The outer protective coat of a **seed** formed from the **integuments** of the **ovule**, after **fertilization**. The testa is usually hard and dry and protects the seed from microorganisms and insects.

testcross *See* **backcross**.

testis The principal reproductive organ in male animals, which produces **spermatozoa**. In vertebrates, the paired testes also produce sex hormones. *See also* **fertilization**.

tetrapods Vertebrates with two pairs of **pentadactyl limbs**. Most tetrapods are land-dwelling.

theory A scientific statement based on experiments that verify a **hypothesis**. *See* **scientific method**.

thermoreceptor A **receptor** that is stimulated by changes in temperature. Examples are heat and cold receptors in skin.

thorax **1.** (in vertebrates) The part of the body containing the heart and lungs (chest cavity). In mammals, it is separated from the **abdomen** by the **diaphragm**.
2. (in insects) The part of the body **anterior** to the abdomen.

thyroid gland An **endocrine gland** in the neck region of vertebrates. When stimulated by *thyroid-stimulating hormone (TSH)* from the **pituitary gland** it produces the hormone *thyroxine*, which controls the rate of growth and development in young animals. For example, in tadpoles, thyroxine stimulates **metamorphosis**.

thyroid-stimulating hormone (TSH) *See* **thyroid gland**.

thyroxine *See* **thyroid gland**.

tibia **1.** One of the segments of the insect leg.
2. The **anterior** of the two bones in the lower hindlimb of **tetrapods**. In humans, it is the shinbone. *See* **endoskeleton**.

tissue A group of similar **cells** specialized to perform a specific function in **multicellular** organisms. Examples are muscle, **xylem**.

tissue fluid *See* **lymph**.

toxin A substance secreted by, for example, bacteria, which is harmful to the organism within which the bacteria are living. *See also* **antibodies**.

trachea **1.** (in land vertebrates) The windpipe leading from the **larynx** and carrying air to the **lungs** where it divides into the **bronchi**. The trachea is supported by **cartilage** rings and has a ciliated **epithelium**, which secretes **mucus** and that traps dust and microorganisms.
   2. (in insects) One of the air tubes in a branching system through which air diffuses into the tissues via the **spiracles**. *See* **gas exchange**.

transect A line marked off in an area, to study the types of species in that area, by sampling the organisms at different points along the line. Measurements of **abiotic factors**, for example, light, soil, **pH**, etc., may also be made along the line to discover any relationship between the distribution of particular species and these factors. *See also* **quadrat**.

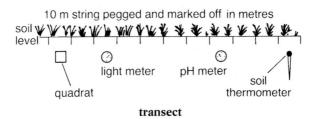

transect

transfer RNA *See* **RNA**, **protein synthesis**.

translocation The transport and circulation of materials within plants: (a) of water and **mineral salts** in **xylem** vessels via the **transpiration stream**; (b) of **carbohydrates** produced by photosynthesis and conducted through the plant in **phloem** sieve tubes.

transpiration The evaporation of water vapour from plant leaves via the **stomata**.

transpiration rate The rate at which water vapour is lost from a leaf to the outside atmosphere. **Transpiration** is affected by several environmental factors:
(a) *temperature*: increased temperature increases water evaporation and thus increases transpiration;
(b) *humidity* (water content of air): increased humidity causes the atmosphere to become saturated with water, thus reducing transpiration;

(c) *wind*: increased air movements accelerate transpiration by preventing the atmosphere around **stomata** from becoming saturated with water. Thus the transpiration rate will be greatest in warm, dry, windy conditions. If the rate of water loss by transpiration exceeds the rate of water uptake, **wilting** may occur.

transpiration stream The flow of water through a plant resulting from **transpiration**. Water evaporates through the **stomata**, causing more water to be drawn by **osmosis** from adjacent leaf cells (**spongy mesophyll** cells).
  The osmotic forces thus set up eventually cause water to be withdrawn from **xylem** vessels in the leaf, resulting in water being pulled through the xylem vessels from the stem and roots, i.e. water evaporation from leaves causes the flow of water (and **mineral salts**) throughout the plant.

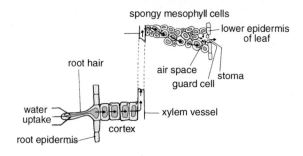

transpiration stream

trophic level The level in a **food chain** at which a group of organisms occurs. Green plants (**food producers**) are at the lowest level and tertiary consumers (**predators**) at the highest level.

tropism A plant growth movement that occurs in response to a stimulus such as light.
  Such movements are related to the direction of the stimulus, the plant organ involved growing either toward or away from it. Tropisms are named by adding a prefix that relates to the stimulus.
  Examples of tropisms are:
(a) **geotropism**: response to gravity;
(b) **phototropism**: response to light;
(c) **chemotropism**: response to chemicals;
(d) **hydrotropism**: response to water.
Tropisms can be either positive or negative depending on whether the response is toward the stimulus or away from it. *See also* **nastic movement**.

Tropisms are important because they cause plants to grow in such a way that they obtain maximum benefit from the environment, in terms of water, light, etc.

Tropisms are caused by an **auxin**, which accelerates growth by stimulating **cell division** and elongation. Uneven distribution of auxin causes uneven growth and leads to bending. *See also* **tropism mechanism**.

tropism mechanism The method by which **tropisms** are controlled by plant **auxins**. The mechanism can be explained by considering experiments done on plant growth and **phototropism** using growing shoots.

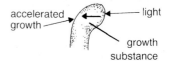

tropism mechanism *A shoot tip.*

A growth substance is produced in shoot tips. This substance diffuses downwards and accelerates growth by stimulating **cell division** and elongation. Uneven distribution of the growth substance results in bending. *See diagram (a).*

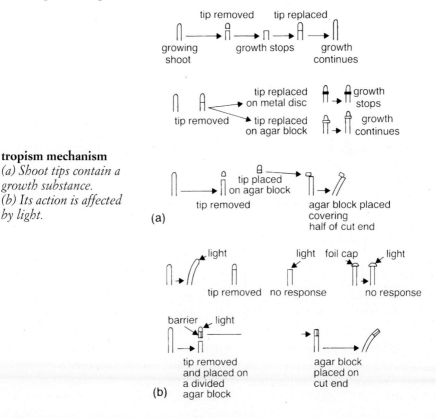

tropism mechanism
(a) Shoot tips contain a growth substance.
(b) Its action is affected by light.

Shoot tips contain cells that are sensitive to light. If light comes from one side only, it causes the growth substance to diffuse away from the illuminated side and accelerate growth at the non-illuminated side. As a result of this uneven growth, the shoot bends towards the light. *See diagram (b).*

trypsin A **protease enzyme** secreted by the vertebrate **pancreas**. *See* **duodenum**.

TSH *See* **thyroid-stimulating hormone**.

tuber An organ of **vegetative reproduction** in flowering plants. Tubers can form from stems or roots. They consist of a food store and buds from which develop new plants.

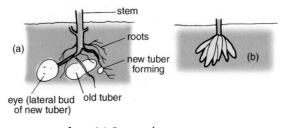

tuber *(a) Stem tuber, e.g. potato.*
(b) Root tuber, e.g. dahlia.

Tullgren funnel An apparatus used to isolate organisms living in the air spaces in soil, e.g. beetles and spiders. The organisms move away from the strong light and high temperature produced by a lamp, and are collected in a jar of preservative.

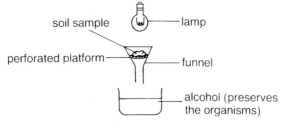

Tullgren funnel

turgor The state of a plant cell after maximum water absorption. Surrounding water enters a cell by osmosis causing the **vacuole** to expand, pushing the **cytoplasm** against the cell wall and making the plant cell solid and strong.

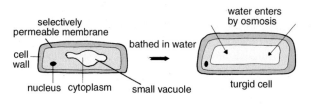

turgor

Turgid cells are important in supporting plants, conferring strength and shape. Young plants depend completely on turgor for support, although in older plants, support is obtained from wood formation (*see* **xylem**).

tympanic membrane *See* **tympanum**.

tympanum or tympanic membrane A thin membrane separating the outer ear and middle ear in tetrapods, i.e. the *eardrum. See* **ear**.

U – V

ulna The **posterior** of the two bones of the lower region of the **tetrapod** forelimb. In humans, it is the larger of the two bones of the forearm. *See* **endoskeleton**.

umbilical cord The cord of blood vessels linking the growing **foetus** in the womb to the **placenta**. It carries nourishment to the foetus and waste products from it. *See also* **birth, pregnancy**.

unicellular (of an organism) Consisting of one **cell** only. Unicellular organisms include **protozoans**, **bacteria** and some **algae**. *Compare* **multicellular**.

urea The main nitrogenous excretory product of mammals. Urea is produced in the **liver** from the **deamination** of excess **amino acids**, and is then excreted by the **kidneys**.

$$H_2N-C-NH_2$$
$$\overset{\|}{O}$$

urea *The chemical structure of urea.*

ureter The tube, in vertebrates, that carries **urine** from the **kidney** to the **bladder**.

urethra The tube in mammals that conveys **urine** from the **bladder** to the exterior. In male mammals, it also serves as a channel for the exit of **spermatazoa**. *See* **kidney, fertilization**.

uterus or womb A muscular cavity in most female mammals that contains the **embryo(s)** during development. The uterus receives **ova** from the **oviduct** and connects to the exterior via the **vagina**. *See also* **fertilization**.

urine A solution of **urea** and **mineral salts** in water produced by the mammalian **kidney**. It is stored in the **bladder** before discharge via the **urethra**.

vaccine A small quantity of **antigens** that is injected into the body. This stimulates the production of the appropriate **antibodies** against a particular **pathogen**, which are then present and available to act if and when that pathogen enters the body.

Vaccines are produced in the following ways:

(a) by separating antigens from the microorganisms (vaccine against influenza);

(b) by mass production of the antigen by **genetic engineering** (vaccine against the hepatitis B virus);

(c) by using the killed pathogen (vaccine against whooping cough);

(d) by chemically changing a **toxin** so that it is no longer toxic but still resembles the antigen (vaccine against tetanus);

(e) by using a live but non-pathogenic strain of the organism (vaccine against rubella).

vacuole A fluid-filled space within a cell **cytoplasm**. containing many compounds such as sugars, in solution. Vacuoles are particularly important in maintaining **turgor** in plant cells. *See also* **cell**, **contractile vacuole**.

vagina A duct in most female mammals that connects the **uterus** with the exterior. It receives the **penis** during **copulation**, and is the route by which the **foetus** is passed during birth. *See* **fertilization**.

valves Membranous structures within the heart and veins in animal **circulatory systems**, which allow blood to flow in one direction only. The heart valves are the following:

(a) *mitral valve* (or *bicuspid valve*): two flaps between the left **atrium** and left **ventricle** of the heart in birds and mammals;

(b) *tricuspid valve*: three flaps between the right atrium and right ventricle of the mammalian heart;

(c) *semilunar valves*: half-moon-shaped flaps in the mammalian heart between the right ventricle and **pulmonary artery**, and the left ventricle and **aorta**. Semilunar valves are also found in the **lymphatic system** and veins. *See also* **heartbeat**.

variation Differences in characteristics between members of the same **species**. There are two main types:

(a) *continuous variations* in which there are degrees of variation throughout the **population** showing **normal distribution** around a mean. For example, in humans: height, weight, pulse rate;

(b) *discontinuous variations* are absolutely clear cut, i.e. there are no intermediate forms; for example, blood groups in humans. Discontinuous variations do not show normal distribution and are used when doing **genetic** crosses.

Variation within a species results either from inherited or environmental factors or a combination of both. Thus, a human being inherits **genes** influencing height, for example, but will also be subject to environmental factors such as nutrition. Inherited variations are considered to be the basis of **evolution** by **natural selection**.

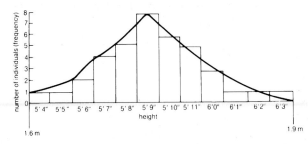

variation *Height in men – an example of continuous variation.*

vascular bundle A strand of longitudinal conducting tissue within plants, consisting mainly of **xylem** and **phloem**. *See also* **root, stem, leaf.**

vasoconstriction *See* **temperature regulation.**

vasodilation *See* **temperature regulation.**

vegetative reproduction or vegetative propagation **Asexual reproduction** in plants by an outgrowth from a parent organism of a **multicellular** body that may become detached and develop independently into a new plant. *See also* **bulb, corm, rhizome, stolon, tuber.**

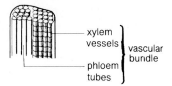

vascular bundle *A vascular bundle in a stem.*

vein 1. A **vascular bundle** in a plant leaf.
2. A **blood vessel** that transports blood from the tissues to the heart. In mammals, veins carry *deoxygenated blood*, i.e. blood that has delivered oxygen to the body (for an exception to this rule, *see* **pulmonary vessels**). Veins connect with smaller vessels called *venules*, which carry blood from the **capillaries**. Veins are thin-walled, and since the **blood pressure** in veins is less than in **arteries**, they have **valves** to prevent the blood flowing away from the heart.

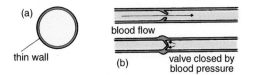

vein *(a) Section through a vein. (b) Valve operation in veins.*

vena cava The largest **vein** in the **circulatory system** of vertebrates. In mammals, it is either of the two main veins:

(a) *superior* (or *anterior*) *vena cava* carries blood from the head, neck and upper limbs into the right **atrium** of the heart;

(b) *inferior* (or *posterior*) *vena cava* carries blood from the rest of the body and lower limbs into the right atrium.

ventral Relating to features of, on, or near that surface of an organism that is normally directed downward, although in humans, it is directed forward. *Compare* **dorsal**.

ventricle *See* **heart**, **heartbeat**.

venule *See* **vein**.

vertebral column or **backbone** A series of closely arranged bones (vertebrae) and/or **cartilages** that runs dorsally from the skull to the tail in vertebrates. It is the principal longitudinal supporting structure and encloses and protects the **spinal cord**. *See* **endoskeleton**.

villi (*singular* **villus**) **1.** Finger-like projections in the vertebrate **intestine** where they occur in large numbers to increase the surface area available for **absorption** of food. *See also* **ileum**.
2. Finger-like projections that develop from the mammalian **placenta** into the **uterus** wall and thus increase the area of contact between maternal and embryonic tissues.

virus The smallest known living particle, having a diameter between 0.025 and 0.25 microns. Viruses are **parasites** that infect animals, plants and bacteria. Virus infections of man include measles, polio and influenza. A virus particle consists of a protein coat surrounding a length of **nucleic acid**, either **DNA** or **RNA**.

bacterium virus becomes attached to bacterium	the virus nuclear material is injected into the bacterium and causes the asembly of new virus parts	the bacterium cell wall is ruptured, releasing many new viruses

virus *A virus infecting a bacterium.*

viscera A collective term for the internal **organs** of an animal.

vitamins **Organic compounds** required in small quantities by living organisms. Like enzymes, vitamins play a vital role in chemical reactions within the body, often regulating an enzyme's action. Shortage of vitamins from the human diet leads to deficiency diseases. The properties of some important vitamins are summarized in the table.

Vitamin	Rich sources	Effects of deficiency
Vitamin A	milk, liver, butter, fresh vegetables	night blindness, retarded growth
Vitamin B_1	yeast, liver	Beri-beri: loss of appetite and weakness
Vitamin B_2	yeast, milk	pellagra: skin infections, weakness, mental illness
Vitamin C	citrus fruits, fresh green vegetables	scurvy: bleeding gums, loose teeth, weakness
Vitamin D	eggs, cod liver oil	rickets: abnormal bone formation
Vitamin E	fresh green vegetables, milk	thought to affect reproductive ability
Vitamin K	fresh vegetables	blood clotting impaired

vitreous humour A transparent jelly-like material that fills the cavity behind the **lens** of the vertebrate **eye**.

voluntary muscle or striated muscle **Muscle** connected to the mammalian skeleton, and which is under the conscious control of the organism; examples are the limb muscles, muscles of face and mouth, etc.

Voluntary muscles involved in limb movement are attached to bones by **tendons** and cause movement by contracting and thus pulling on bones, particularly at joints. *See also* **antagonistic muscles**, **involuntary muscle**.

W

wall pressure The resistance to stretching in a plant cell. A cell absorbing water by **osmosis** will continue to expand until its **selectively permeable membrane** and cell wall can stretch no further. Wall pressure increases as the point of **turgor** is approached.

warm blooded *See* **endotherm**.

water A compound consisting of the elements hydrogen and oxygen. Its chemical formula is H_2O. Water can be synthesized by burning hydrogen in air.

$$\text{hydrogen} + \underset{\text{(oxygen)}}{\text{air}} \xrightarrow{\frac{\text{heat}}{\text{energy}}} \text{water}$$

Water can be split into its constituent elements:

$$\text{water} \xrightarrow{\frac{\text{electrical}}{\text{energy}}} \text{hydrogen} + \text{oxygen}$$

$$\text{water} \xrightarrow{\frac{\text{light}}{\text{energy}}} \text{hydrogen} + \text{oxygen}$$

The latter reaction occurs in green plants during photosynthesis.

Water is a colourless, tasteless compound that freezes at $0\,°C$ and boils at $100\,°C$. Water can exist in several states (the *water cycle*) as shown in the diagram.

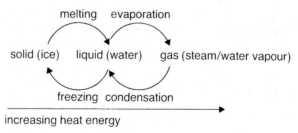

water *The water cycle.*

Water is biologically important for the following reasons:
(a) Water is a major constituent of cells. Water is the most abundant component of organisms. It is estimated that the human body is more than 60% water. The water content of living organisms can be estimated by weighing food samples, e.g. a potato, drying the samples in an oven, and then reweighing. The resulting loss of weight gives the water content. The water content of living organisms is usually between

60% and 95%, although it can be as high as 99% (as in jellyfish) and as low as 20% (in plant seeds).

(b) Water is a solvent. Water is called 'the universal solvent' since more substances dissolve in water than in any other liquid. This is important, since all the chemical reactions that occur in organisms take place in *aqueous solution*, i.e. dissolved in water (*aqua* = water).

(c) Water is a means of transport. Within organisms, substances such as food are required to be transported throughout the organism. Such materials are transported in aqueous solution, e.g. in blood or **phloem**.

(d) Water is important in syntheses. The synthesis of many compounds in living organisms requires water. For example, one of the raw materials for photosynthesis is water.

(e) Water is required in lubrication. The internal organs and joints of living organisms must be lubricated to prevent friction during movement. Various lubricating fluids exist, e.g. **mucus**, all of which are aqueous solutions.

(f) Water is important in reproduction. Many organisms use water to transport the male reproductive cells (**sperm**) to the female cells (**ovum**) so that fertilization can occur.

(g) Water is important in temperature regulation. The evaporation of water (e.g. sweat) from the surface of organisms has a cooling effect, while the high water-content of cells provides insulation and prevents rapid temperature changes.

water purification The treatment applied to tap water in order to make it safe to drink. The water is filtered through sand and gravel to remove large particles and then chlorine is added to kill harmful microorganisms.

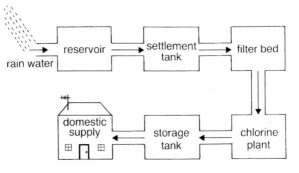

water purification

weathering The process by which exposed rock is converted into soil. The rock is broken down to small particles by the effects of wind, rain, heat and frost. These rock particles form the 'skeleton' of the soil but are only one component of a complex mixture of inorganic and organic factors that make up a soil.

white blood cell, white blood corpuscle or **leucocyte** One of various types of **blood** cell found in most vertebrates. Their function is in defence against microorganism infection, which they achieve by **phagocytosis** or by **antibody** production.

white blood corpuscle *See* **white blood cell**.

wild type An organism having a **phenotype** or **genotype** that is characteristic of the majority of the **species** in natural conditions.

wilting A plant condition occurring when water loss by **transpiration** exceeds water uptake. The cells lose **turgor** and the plant droops.

womb *See* **uterus**.

wood *See* **xylem**.

X – Y – Z

xylem Tissue within plants that conducts water and **mineral salts**, absorbed by roots from the soil, throughout the plant. Xylem tissue consists of long continuous tubes formed from columns of cells in which the horizontal cross-walls have disintegrated and the cell contents have died.

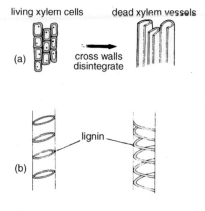

The vessels thus formed are strengthened by a compound called **lignin**, and ultimately form the *wood* of the plant. Associated with xylem vessels, and providing additional strength, are specialized fibrous cells called *xylem fibres*, some of which are useful to humans, for example, flax. Thus xylem is commercially important as a source of wood and fibres. *See also* **leaf, root, stem.**

xylem (a) Long, continuous vessels are formed as horizontal cross-walls disintegrate. (b) The vessels are strengthened by lignin.

yeast A unicellular **fungus** that is important in baking, **brewing** and **fermentation**. *See* **budding**.

yellow spot *See* **fovea**.

yolk A store of food material, mainly protein and fat present in the eggs (**ova**) of most animals. In fish, reptiles and birds, the yolk is contained within a *yolk sac*, which is absorbed into the **embryo** as the yolk is used.

zooplankton *See* **plankton**.

zygote The **diploid** cell resulting from the fusion of two **gametes** during **fertilization**. The tiny ball of cells, formed by cleavage of the zygote (**mitosis**), becomes embedded in the wall of the **uterus** in mammals (**implantation**). *See also* **fertilization, pregnancy**.

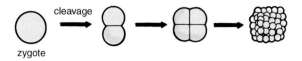

zygote *The zygote divides by mitosis after fertilization, to form a ball of cells.*

Appendix A

Chemical elements

In this table, the names and symbols of the chemical elements are given, with their proton numbers (Z), numbers of isotopes (n_i) and melting and boiling temperatures (T_m, T_b).

Element	Z	n_i	$T_m/°C$	$T_b/°C$
Actinium Ac	89	7	1230	3100
Aluminium Al	13	7	660	2400
Americium Am	95	8	1000	2600
Antimony Sb	51	18	631	1440
Argon Ar	18	7	−190	−186
Arsenic As	33	11	–	610
Astatine At	85	7	250	350
Barium Ba	56	16	710	1600
Berkelium Bk	97	6		
Beryllium Be	4	4	1280	2500
Bismuth Bi	83	12	271	1500
Boron B	5	4	2030	3700
Bromine Br	35	18	−7	58
Cadmium Cd	48	18	321	767
Caesium Cs	55	15	27	690
Calcium Ca	20	11	850	1450
Californium Cf	98	7		
Carbon C	6	6	3500	3900
Cerium Ce	58	13	804	2900
Chlorine Cl	17	10	−101	−34
Chromium Cr	24	8	1900	2600
Cobalt Co	27	10	1490	2900
Copper Cu	29	10	1080	2580
Curium Cm	96	7	1340	
Dysprosium Dy	966	12	1500	2300
Einsteinium Es	99	10		
Erbium Er	68	10	1530	2600
Europium Eu	63	12	830	1450
Fermium Fm	100	7		
Fluorine F	9	4	−220	−188
Francium Fr	87	5	30	650
Gadolinium Gd	64	14	1320	2700

Element	Z	n_i	$T_m/°C$	$T_b/°C$
Gallium Ga	31	10	30	2250
Germanium Ge	32	13	960	2850
Gold Au	79	13	1060	2660
Hafnium Hf	72	11	2000	5300
Helium He	2	3	–	–269
Holmium Ho	67	6	1500	2300
Hydrogen H	1	3	–259	–253
Indium In	49	19	160	2000
Iodine I	52	17	114	183
Iridium Ir	77	10	2440	4550
Iron Fe	26	8	1539	2800
Krypton Kr	36	19	–157	
Lanthanum La	57	8	920	3400
Lawrencium Lw	103	1		
Lead Pb	82	24	327	1750
Lithium Li	3	4	180	1330
Lutetium Lu	71	5	1700	3300
Magnesium Mg	12	6	650	1100
Manganese Mn	25	9	1250	2100
Mendelevium Md	101	1		
Mercury Hg	80	16	–39	357
Molybdenum Mo	42	15	2600	4600
Neodymium Nd	60	13	1020	3100
Neon Ne	10	7	–250	–246
Neptunium Np	93	8	640	3900
Nickel Ni	28	11	1450	2800
Niobium Nb	41	15	2400	5100
Nitrogen N	7	6	–210	–196
Nobelium No	102	1		
Osmium Os	76	13	3000	4600
Oxygen 0	8	6	–219	–183
Palladium Pd	46	17	1550	3200
Phosphorous P	15	7	44	280
Platinum Pt	78	12	1770	3800
Plutonium Pu	94	11	640	3500
Polonium Po	84	12	250	960
Potassium K	19	8	63	760
Praseodymium Pr	59	8	930	3000

Element	Z	n_i	$T_m/°C$	$T_b/°C$
Promethium Pm	61	8	1000	1700
Protactinium Pa	91	9	1200	4000
Radium Ra	88	8	700	1140
Radon Rn	86	7	−71	−62
Rhenium Re	75	7	3180	5600
Rhodium Rh	45	14	1960	3700
Rubidium Rb	37	16	39	710
Ruthenium Ru	44	12	2300	4100
Samarium Sm	62	14	1050	1600
Scandium Sc	21	11	1400	2500
Selenium Se	34	16	220	690
Silicon Si	14	6	1410	2500
Silver Ag	47	16	960	2200
Sodium Na	11	6	98	880
Strontium Sr	38	13	77	1450
sulphur S	16	7	119	445
Tantalum Ta	73	11	3000	5500
Technetium Tc	43	14	2100	4600
Tellurium Te	52	22	450	1000
Terbium Tb	65	8	1360	2500
Thalium Tl	81	16	300	1460
Thorium Th	90	9	1700	4200
Thulium Tm	69	10	1600	2100
Tin Sn	50	21	231	2600
Titanium Ti	22	8	1680	3300
Tungsten W	74	10	3380	5500
Uranium U	92	12	1130	3800
Vanadium V	23	7	1920	3400
Xenon Xe	54	22	−111	−108
Ytterbium Yb	39	12	1500	1500
Yttrium Y	70	11	820	3000
Zinc Zn	30	13	420	907
Zirconium Zr	40	12	1850	4400

Appendix B

Units of measurement

Length

1 metre (m)= 100 centimetres (cm)
1 centimetre= 10 millimetres (mm)
1 millimetre= 1000 microns (gm)
1 micron= 1000 nanometres (nm)

Volume

1 litre (l)= 1000 cm^3 (millilitres (ml))

Mass

1 tonne=1000 kilogrammes (kg)
1 kilogramme=1000 grammes (g)

Temperature

boiling point of water= 1000 Celsius (°C) freezing point of water=0°C
normal average human body temperature=37°C

Energy

1 kilojoule (kJ)= 1000 Joules (J)=240 calories (C)

Food type	Energy value
Carbohydrate	17 kJ/g
Protein	17 kJ/g
Fat	39 kJ/g

Appendix C

Characteristics of living things

For an organism to be considered as living it must demonstrate all of the following features:

Movement   The ability to change position either of all, or part of the body.

Excretion   The ability to remove from the body waste materials produced by the organism.

Respiration   The ability to release energy by the breakdown of complex chemicals.

Reproduction   The ability to produce offspring.

Irritability   The ability to sense and respond to changes in the environment.

Nutrition   The ability to take in or manufacture food that can be used when required as a source of energy or as building materials.

Growth   The ability to increase in size and complexity through the production of new cell material.

Appendix D

The differences between plants and animals

Plants	Animals
Cell surrounded by cellulose cell wall	No cellulose cell wall
Large vacuoles in cells with cell sap	Vacuoles when present only small
Large cells with definite shape	Small irregularly shaped cells
Only restricted movement possible	Free movement possible
Response to stimulus slow	Rapid response to stimulus
Cells contain chloroplasts (chlorophyll)	No chloroplasts (chlorophyll)
Photosynthesize	Must obtain food from external sources

These characteristic differences should only be regarded as guidelines. Attempts to classify certain organisms (for example, bacteria, fungi, viruses) within these terms of reference will be difficult and have provided scientists with great problems.

Appendix E

The major groups of living organisms
THE ANIMAL KINGDOM (MAJOR PHYLA)

(a) INVERTEBRATES: Animals without a **vertebral column** (backbone)

Phylum Protozoa: Microscopic **unicellular** animals.
Phylum Porifera: Porous animals often occurring in colonies, for example, sponges.

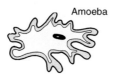
Amoeba

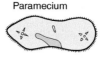
Paramecium

bath sponge

Phylum
coelenterata:
Tentacle-bearing animals with stinging cells.

Hydra   Jelly fish   Sea anenome

Phylum Platyhelminthes: Flatworms.

Planaria

Tapeworm

Phylum Annelida: Segmented worms.

Earthworm   Sandworm

Leech

Phylum Mollusca: Soft-bodied animals often with shells.
Phylum Arthropoda: Jointed limbs: *exoskeleton*.

Phylum Echinodermata: Spiny-skinned marine animals.

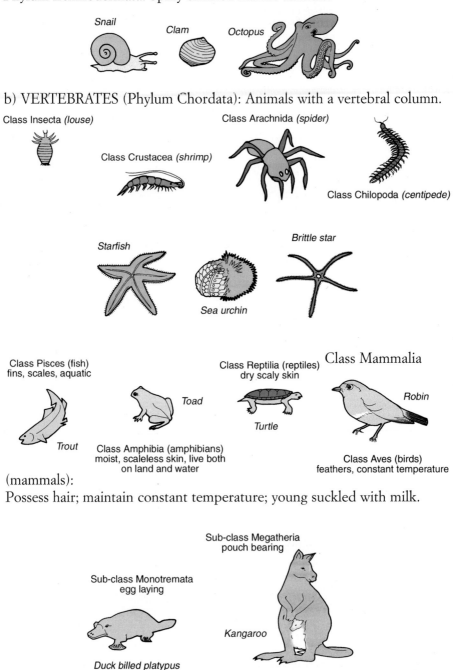

b) VERTEBRATES (Phylum Chordata): Animals with a vertebral column.

Snail

Clam

Octopus

Class Insecta *(louse)*

Class Arachnida *(spider)*

Class Crustacea *(shrimp)*

Class Chilopoda *(centipede)*

Starfish

Brittle star

Sea urchin

Class Pisces (fish)
fins, scales, aquatic

Class Reptilia (reptiles)
dry scaly skin

Class Mammalia

Toad

Robin

Trout

Turtle

Class Amphibia (amphibians)
moist, scaleless skin, live both
on land and water

Class Aves (birds)
feathers, constant temperature

(mammals):
Possess hair; maintain constant temperature; young suckled with milk.

Sub-class Megatheria
pouch bearing

Sub-class Monotremata
egg laying

Kangaroo

Duck billed platypus

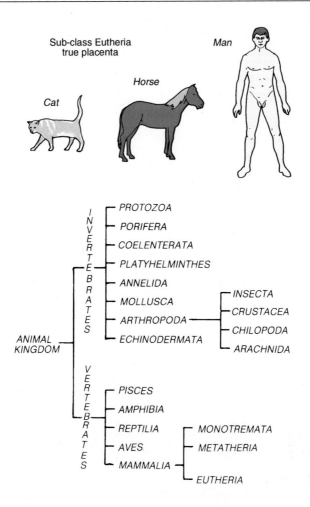

Sub-class Eutheria
true placenta

Man

Horse

Cat

THE PLANT KINGDOM (MAJOR PHYLA)

Phylum Thallophyta: Unicellular and simple multicellular plants.

Class Algae: Photsynthetic; includes unicellular, filamentous, and multicellular types.

Class Fungi: **Heterotrophic**; including both **parasites** and **saprophytes**.

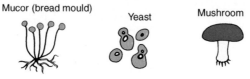

Phylum Bryophyta: Green plants with simple leaves and showing alternation of generations; moist **habitats**.

Class Hepaticae (liverworts)

Phylum Pteridophyta (ferns; bracken; horsetails): Green plants, with **roots**, **stems**, leaves, and showing alternation of generations.

Phylum Spermatophyta: Plants that produce seeds.

Class Gymnospermae: Seeds produced in **cones**.

Class Angiospermae: Flowering plants; seeds enclosed within fruits.

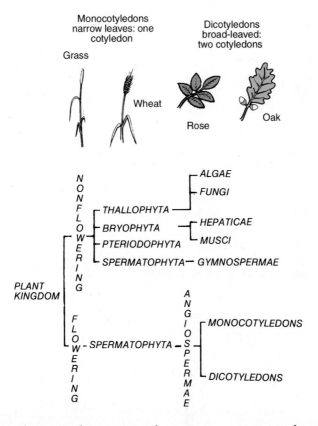

Bacteria and viruses do not meet the criteria necessary to be placed in either the animal or plant kingdoms.

Appendix F

How biologists look at things

Terms such as **dorsal**, **anterior**, **posterior** relate to a particular surface of an organism. The location of these areas depends on the orientation of the organism. For example, in the dog, the ventral surface points downwards, whereas in man, it points forwards.

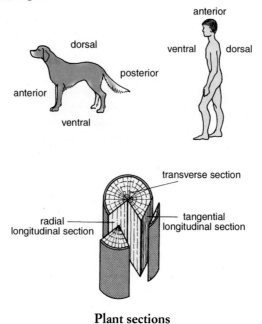

Plant sections

Appendix G

Your aim should be to give an account which would enable another scientist to copy precisely your procedure and to draw conclusions from your observations.

The traditional way of reporting an experiment includes:

Title	A summary of the aim or aims.
Introduction	A brief outline of useful background information.
Apparatus	A complete list (or labelled diagram) of everything used.
Method	An ordered account of what was done. (A labelled diagram may help.)
Result	A record of observations and measurement using tables, graphs, diagrams or description.
Discussion and Conclusion	An objective account of what has been learned, problems encountered, and further investigations that arise from the observations.

Appendix H

A diagram represents something as simply as possible. You do not have to be an artist to draw one, but you do need to follow a few basic rules:

• Equip yourself with a sharp pencil, a ruler and a rubber.
• Draw diagrams large enough to show detail easily.
• Lines should be single and complete to give a neat outline.
• Lines should complete a structure without leaving holes, going too far or blocking passages.
• The use of colours or shading should only be used to make an important feature stand out.
• Diagrams require a clear title.
• Labels are used to indicate the names of the parts shown.
• Annotations are used to give a brief description of the parts shown.
• Labels and annotations should be arranged horizontally around the outside of the diagram.
• Lines (drawn with a ruler) should precisely connect the labels and annotations to the parts to which they refer.
• Label and annotation lines should never cross each other.

A good diagram is often worth more than a long description.

Appendix I

Presentation of information
(pie charts, line graphs, bar charts, averages, ratios, percentages)

It is often helpful to present information in the form of a chart or graph as it makes the information easier to understand. Three common ways of doing this are shown below.

Pie chart

This is a diagram in the form of a circle (the 'pie') divided into segments ('slices') to show the proportions of the quantities to be compared.

For example:

The following information was gathered about the various eye colours of pupils in a school:

eye colour	number of pupils
blue	200
brown	100
grey	50
green	50

This information can be presented as a pie chart. Your pie chart must have a title and a *legend* identifying each of the segments. You convert the numerical information to visual form as follows:

The total number of pupils is 400, represented by the whole circle (100%):

200 have blue eyes and take up one half of the circle
  (200 is one half (or 50%) of 400)
100 have brown eyes and take up one quarter of the circle
  (100 is one quarter (or 25%) of 400)
50 have grey eyes and take up one eighth of the circle
  (50 is one eighth (or 12.5%) of 400)
50 have green eyes and take up one eighth of the circle
  (50 is one eighth (or 12.5%) of 400).

Line graph

Where two quantities are measured or counted in relation to one another, a line graph is used. For example, the following temperatures and times were noted using a thermometer and stopclock as some water cooled down in a plastic beaker:

time (min)	temperature (°C)
0	100
10	70

20   50
30   40
40   35

These numerical measurements can be shown on a line graph. Remember to give your line graph a title. Each axis must also be given a scale and a label, and the units of measurement must be shown.

Bar charts

The bar chart, like the pie chart, can be used to show relative quantities. For example, the following information was gathered about the trees in a Forestry Commission wood in Scotland:

type of tree	number of trees
spruce	90
fir	55
birch	70
rowan	30
ash	50

This numerical information can be converted into separate bars, the different heights corresponding to the numbers on a vertical scale. Your bar chart must therefore be given a scale, which must be identified with a label. Each bar must also have its own label. Remember to give your bar chart a title.

Averages

A scientist often repeats an experiment, or takes several measurements of the same thing. This is to make sure that the results can be trusted. If you have several sets of results, they are not likely to be all the same. Rather than choose just one of these results, it is fairer to take an average of the set of results.

It is not difficult to work out an average. There are two steps:
1. add up all the numbers in the set of results, to get a total

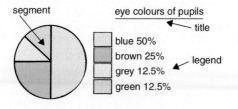

segment   eye colours of pupils
                                    ← title

blue 50%
brown 25%
grey 12.5%   ← legend
green 12.5%

2. divide this total by the number of measurements or results in the set. For example: find the average of the following times:

9, 11, 15, 23, 32 minutes

1. find the total:

9
11
15
23
32
90 minutes

2. divide the total by the number of measurements (in this case, 5):

90 ÷ 5 = 18

Therefore the average time is 18 minutes.

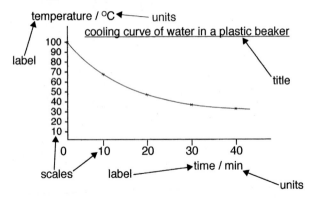

Ratios

Relative proportions can be expressed numerically as a ratio. The ratio of a to b, where a and b represent whole numbers, is

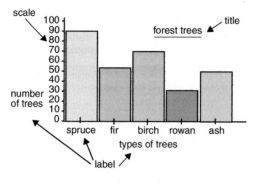

$$a \div b \text{ or } a/b$$
and is often written $a{:}b$.

For example, if the number of pupils in a biology class is 20, and 2 are absent, the the ratio of the number absent to the number present is

$$1 \div 9 \text{ or } \tfrac{1}{9} \text{ or } 1{:}9.$$

Normally, ratios are expressed in the simplest whole numbers.
For example:

A food sample was tested and found to contain 4 g of fat and 28 g of carbohydrate. Calculate the ratio of carbohydrate to fat in simple whole numbers.

$$\text{28 g carbohydrate:4 g fat}$$
$$28{:}4$$
$$7{:}1$$

Percentages

It is often very useful to express fractions so that their denominators are 100. A fraction with a denominator of 100 is called a *percentage*:
$^{75}/_{100} = 75$ per cent $= 75\%$
Per cent means per hundred.
To express a fraction as a percentage:
If the total mass of the food sample above is 50 g, the percentage of fat is
$^{4}/_{50} \times 100 = 8\%$
and the percentage of carbohydrate is $^{28}/_{50} \times 100 = 56\%$.

Appendix J

Designing and carrying out investigations

Investigatory experimental work is a fundamental aspect of all biology courses. You will be assessed in three skills:

1. designing investigations
2. carrying out investigations
3. interpreting investigations.

When doing your investigations, use the following guidelines, which show the ten levels of attainment which are normally assessed.

Level of attainment	Designing	Carrying out	Interpreting
1	Ask questions about familiar events.	Note observable features of familiar events.	Discuss your observations.
2	Ask questions 'how?' 'why?' 'what will happen if?' Suggest ideas. Make predictions.	Make a series of related observations.	Use your observations to support conclusions. Compare your observations with what you expected.
3	Suggest ideas, ask questions and make predictions about an event and show how they could be tested.	Make careful measurements and observations using instruments.	Be aware that your conclusions may be invalid unless a fair test has been done. Give a simple explanation of your observations.
4	Suggest ideas, ask questions and make predictions that can be tested, based on some prior knowledge.	Carry out a fair test, using suitable apparatus, and take appropriate measurements, e.g. temperatures; volumes.	Identify any pattern in your results. Draw conclusions which relate the pattern to your orignal predictions.
5	Formulate a hypothesis which links the independent and dependent variables, based on scientific knowledge. Suggest a method to test your hypothesis.	Carry out a fair test, taking a range of measurements which produce results to test the relationship between the variables.	Draw conclusions which link the independent and dependent variables. Consider the validity of your conclusions.

Level of attainment	Designing	Carrying out	Interpreting
6	Suggest a relationship between two continuous variables using scientific knowledge, and how you would investigate that relationship.	Carry out observations and measurements which would give fine discrimination, using appropriate apparatus.	Interpret your results to explain the relationship between the two continuous variables, in terms of a suitable model.
7	Predict the relative effects of two or more variables under investigation. Using scientific knowledge, suggest how to investigate your predictions.	Carry out an investigation in which you control the appropriate variables, and measure their relative effects.	Interpret your results in terms of the relative effects of the independent variables. Be aware of, and explain, the limitations of your evidence.
8	Generate quantitative predictions using scientific knowledge, and plan an investigation to investigate your predictions.	Carry out an investigation using measuring instruments which will provide very accurate quantitative results.	Consider to what extent your results support your predictions, and evaluate the contribution of each part of the investigation to the overall conclusion.
9	Generate quantitative predictions, and form a hypothesis, based on a scientific theory. Suggest an appropriate range of investigatory techniques.	Systematically use a range of practical investigatory techniques to obtain sufficient data to judge the relative effects of the factors involved in your predictions.	Analyse and interpret your data in terms of your quantitative predictions and be aware of the uncertainty of the evidence collected.
10	Develop hypotheses to explain events you have studied using scientific knowledge, laws, theories and models. Plan extended practical investigation to test your hypotheses.	Use a range of investigatory techniques to collect reliable and valid data which will allow you to make a critical evaluation of the law, theory or model being tested.	Analyse and use your data to evaluate the law, theory or model in terms of how it can explain the observed events.